MINI FARMING
SELF SUFFICIENCY GUIDE
FARMING ON AN ACRE

The Complete Mini Farming Guide on an Acre or Smaller Area From Starting and Maintaining a Homestead Intensive Farms, Minimizing Resources and Maximizing Profit.

Table of Contents

Introduction

Intensive Agriculture, in Agricultural Economics, is system of cultivation using large amounts of labour and capital relative to land area. Large amounts of labour and capital are necessary to the application of fertilizer, insecticides, fungicides, and herbicides to growing crops, and capital is particularly important to the acquisition and maintenance of high-efficiency machinery for planting, cultivating, and harvesting, as well as irrigation equipment where that is required.

Optimal use of these materials and machines produces significantly greater crop yields per unit of land than extensive agriculture, which uses little capital or labour. As a result, a farm using intensive agriculture will require less land than an extensive agriculture farm to produce a similar profit. In practice, however, the increased economies and efficiencies of intensive agriculture often encourage farm operators to work very large tracts in order to keep their capital investments in machinery productively engaged — i.e., busy.

On the level of theory, the increased productivity of intensive agriculture enables the farmer to use a relatively smaller land area that is located close to market, where land values are high relative to labour and capital, and this is true in many parts of the world. If costs of labour and capital outlays for machinery and chemicals, and costs of storage (where desired or needed) and transportation to market are too high then farmers may find it more profitable to turn to extensive agriculture.

However, in practice many relatively small-scale farmers employ some combination of intensive and extensive agriculture, and many of these operate relatively close to markets. Many large-scale farm operators, especially in such relatively vast and agriculturally advanced nations as Canada and the United States, practice intensive agriculture in areas where land values are relatively low, and at great distances from markets, and farm enormous tracts of land with high yields. However, in such societies overproduction (beyond market demands) often results in diminished profit as a result of depressed prices.

Though the term intensive agriculture elicits images of vast tracts of land, monoculture, pesticides, and barns filled with thousands of animals hardly able to stand, this is not how the practice started. Intensive farming originated in the ancient civilizations of Egypt, Mesopotamia, India, Pakistan, North China, Mesoamerica, and Western South America with the creation of water management systems and the domestication of large animals that could pull plows. In more recent years, and especially since industrialization, intensive agriculture has also come to be characterized by a variety of other practices such as heavy pesticide use, rotational grazing, and concentrated animal feeding operations (CAFOs).

What Is Intensive Agriculture?

Intensive agriculture is a method of farming that uses large amounts of labor and investment to increase the yield of the land. In an industrialized society this typically means the use of pesticides, fertilizers, and other chemicals that boost yield, and the acquisition and use of machinery to aid planting, chemical application, and picking. In theory, this reduces the amount of land needed for an economically viable farm to grow crops or raise animals. However, in countries such as the United States and Canada these methods are often used to overproduce products as companies attempt to increase their market share. Profit is then diminished so that farmers must continue overproducing in order to stay economically viable, and often seek compensation for low profits via government subsidies.

What Are the Characteristics of Intensive Agriculture?

- **Pasture Intensification**

Pasture intensification is the increase in value and production that occurs due to inputs such as money, labor, and pesticides, specifically in the pastures on which farmed animals graze. Historians believe that pasture intensification, and agricultural intensification more broadly, was a necessary step in creating the modern societies we have today, as new methods of farming and increasing yield allowed for larger populations to grow.

The most common and effective way of increasing inputs throughout history has been to plant or graze more land, leading to an increase in the yield of the farm. Simply increasing the amount of land being used, however, can have serious consequences for biodiversity, which is lost when native plants and grasses are cleared to make room for grazing.

In recent years there has been an increased level of interest in methods of intensification that reduce some of its negative effects. These include cultivating certain crops, such as soybeans, in pastures that cows graze on.

- **Rotational Grazing**

Rotational grazing is a type of pasture intensification that entails breaking grazing areas into smaller paddocks. Farmed animals are rotated through the different paddocks one by one, allowing those not in use to recuperate and regrow foliage. This is distinct from traditional grazing, as typically cattle are allowed to free graze an entire pasture, which destroys the plant life and does not provide adequate time for regrowth. This leads to more land needing to be used to support the farmed animals.

Concentrated Animal Feeding Operations (CAFOs)

Concentrated animal feeding operations (CAFOs) are the predominant type of animal farm in industrialized systems of agriculture. They consist of large numbers of animals on relatively little land. Instead of grazing and gathering their own food, the animals have food brought to them. The animals are confined in small spaces with little room or opportunity to express natural behaviors.

- **Crop Irrigation**

Crop irrigation is the use of manmade systems to control water application and make up for any shortage of natural rainfall. In California more than nine million acres of land are irrigated, accounting for 80 percent of the total water used by businesses and homes. The heavy use of irrigation to grow crops in areas that are not able to naturally sustain them creates risks and challenges, especially because of the ongoing threat of drought in many such places.

- **Genetically Modified Organism (GMO) Seeds**

Many of the most abundant crops in the United States are species that have been genetically modified. In 2018, 94 percent of all soybeans, 94 percent of cotton, and 92 percent of corn being planted in the country were genetically modified. Generally, seeds and crops are genetically modified to be larger, more pest-resistant, or to tolerate herbicides better.

- **Use of Agrochemicals**

Modern-day industrialized intensive agriculture uses large amounts of pesticides and fertilizers. These chemicals wreak havoc on ecosystems, polluting water and killing off important species such as bees and ladybugs.

Intensive Agriculture Examples

- **Livestock**

Most of the farmed animals in the United States live a significant portion of their lives on industrial factory farms that use a variety of intensive methods to produce more meat, dairy, or eggs for less money. One such method is keeping the animals enclosed in small spaces and delivering their food to them. This forces them to grow more quickly and reduces the need for space. Another example of intensive methods in animal agriculture is the use of selectively bred animals that grow more quickly than naturally occurring breeds and get large enough for slaughter in a shorter period of time. This often results in harsh repercussions for the animals themselves, such as difficulty breathing, walking, and standing.

- **Aquaculture**

Intensive agriculture is apparent in every part of the industry, and aquaculture is no exception. One example is the standard practice of housing extremely high densities of fish in artificial tanks, allowing the farmers to control feed, oxygen levels, and a variety of other factors leading to an increase in yield.

- **Crops**

There are several ways that farmers who grow crops use intensive agriculture to produce higher yields. Tactics include the use of pesticides, insecticides, fertilizers, irrigation, and the use of genetically modified seeds.

Intensive Versus Extensive Agriculture

- **Methodology**

Intensive farming focuses on investing a lot of resources and labor into small tracts of land in order to increase yield. Extensive agriculture, on the other hand, employs larger tracts of land and lower quantities of labor and resources.

- **Location**

Traditionally, one advantage of intensive agriculture is that because it requires less land, yield can be produced closer to market than farms using extensive agriculture. However, most modern farms using intensive methods in higher-income countries operate on a large scale, often on thousands of acres, and in areas far from where consumers live.

- **Farm Land Area**

In theory, extensive farming requires much more land than intensive agriculture, as additional chemicals, machinery, and labor are not being applied to increase yield. However, due to the shift in farming techniques favoring intensive methods for larger tracts of land, both intensive and extensive farming techniques use large amounts of land today.

- **Inputs**

Intensive farming requires greater inputs than extensive farming. Intensive farms tend to use more labor, agrochemicals, and special seeds or breeds of animals. Extensive farming largely relies on the natural fertility of the land and the natural behaviors of the animals.

- **Profitability**

Modern-day intensive farming has sought to produce massive amounts of food as cheaply as possible. This has resulted in the overproduction of many food items, which drives the market price down. For extensive farming to be profitable large amounts of land are required. Because of this, extensive methods tend to be used most frequently where population densities are low and land is inexpensive.

- **Productivity (Yield/Hectare)**

Unsurprisingly, because the goal of intensive agriculture is to maximize the productivity of land, the yield per hectare is higher than that of farms that utilize extensive methods.

- **Environmental Impact**

Both extensive and intensive farming can have negative environmental impacts. Extensive farming requires large amounts of arable land and has often led to deforestation, while intensive farming involves chemicals that negatively impact the environment and native species. Feed for intensively farmed animals is a growing factor in deforestation as well. Growing the food used to feed animals in intensive farming also leads to deforestation in South America and increasingly in other areas of the world.

Why Is Intensive Agriculture Bad?

- **Animal Cruelty**

Billions of animals in the United States suffer on factory farms that employ intensive methods to increase profitability. Often they are confined in such small spaces that they can barely move. Standard procedures include debeaking, castration, tail docking, and dehorning. All of these frequently occur without sedation, causing large amounts of suffering and pain for the animals that endure them.

- **Deforestation**

Because intensive agriculture has shifted from focusing on maximizing the productivity of small pieces of land to application on farms spanning thousands of acres, it can often drive deforestation even before one considers the sources of animal feed. Because the land must be easily accessible for planting, watering, and fertilizing, trees must be removed to create large expanses of flat land. Growing the corn and soy to feed these animals is a leading cause of deforestation globally.

- **Human Health**

Exposure to the pesticides that intensive agriculture tends to use in large quantities can have a number of negative effects on human health. These include irritation to the skin and eyes and negative effects on the nervous and endocrine systems. The mismanagement of the large amounts of manure produced on CAFOs can also lead to health problems in surrounding communities.

- **Pest and Weed Resistance**

Following repeated application of a particular pesticide or herbicide, many pest species, both plant and animal, can build resistance. This often results in stronger chemicals being used to destroy the target species, or a larger amount or higher concentration of the chemicals being applied.

- **Soil Degradation**

Intensive farming contributes to soil degradation, as land tends to be planted on repeatedly without providing a break for the dirt to recover its nutrients. This often results in the increased use of fertilizers to make up for the lack of nutrients in the soil.

- **Water Pollution**

Intensive farming methods contribute considerably to water pollution. Every year the animals on factory farms in the United States produce billions of gallons of waste. With nowhere else to go this waste tends to be stored in large cesspools and be sprayed over fields. Both of these systems of disposal result in water pollution, as the waste sinks into the groundwater or makes its way into rivers, lakes, or other bodies of water.

- **Climate Change**

The increase of intensive farming methods around the world was noted in 2020 as threatening the world's chance of meeting the terms of the Paris agreement. Both the use of artificial fertilizers and the farming of animals — especially cows, who produce large quantities of methane and are often fed with grains farmed on deforested land — are causes of increasing emissions of greenhouse gases.

- **Harm to Smaller Farms**

The rise of intensive agriculture has dealt a serious blow to small farms. Because the larger, corporate enterprises can afford to produce crops and animals on a much larger scale, they are able to sell them for a lower market price while still being able to make a profit. Smaller farms, however, have been left behind, as they do not tend to produce enough to be able to accept such low prices. This has contributed to many farmers leaving the industry and further social effects for farming communities.

The intensification of farming has played an important role in the history of agriculture. It allowed for farmers to feed growing communities around the world. However, intensive agriculture as we know it today is no longer sustainable or necessary. The methods employed have countless negative impacts on the environment, human health, animal lives, and communities, caused by the heavy use of chemicals and the inhumane treatment of animals and workers that are trademarks of modern-day intensive farming. Thankfully, we can take steps to reduce our support of the industry by purchasing locally grown foods from farms that employ more environmentally friendly and ethical methods of production.

How Mini-Farming works for you

Many homeowners undertake the task of gardening or small-scale farming as a hobby to get fresh produce and possibly save money over buying food at the supermarket. Unfortunately, the most common gardening methods end up being so expensive that even some enthusiastic garden authors state outright that gardening should be considered, at best, a break-even affair.

Looking at the most common gardening methods, these authors are absolutely correct. Common gardening methods are considerably more expensive than necessary because they were originally designed to benefit from the economies of scale of corporate agribusiness. When home gardeners try to use these methods on a smaller scale, it's a miracle if they break even over a several-year period, and it is more likely they will lose money.

The Economics of a Mini-Farm

The cost of tillers, watering equipment, large quantities of water, transplants, seeds, fertilizers and insecticides adds up quickly. Balanced against the fact that most home gardeners grow only vegetables, and vegetables make up less than 10 percent of the calories an average person consumes, it quickly becomes apparent that even if the cost of a vegetable garden were zero, the amount of actual money saved in the food bill would be negligible.

For example, if the total economic value of the vegetables collected from the garden in a single season amounted to about $350, even if the vegetables could be produced for free, the economic benefit would amount to only $7 a week when divided over the year.

The solution to this problem is to both cut costs and increase the value of the end product. Using this combination, the economic equation balances in favor of the gardener instead of the garden supply store, and it becomes quite possible to supply all of a family's food except meat (if you eat it) from a relatively small garden. According to the USDA, the average yearly cost to feed a family of three is $8,140.92 — on a low-cost plan. Increase that to a liberal plan and we're looking at an average cost of more than $12,000 a year. Understanding that food is purchased with after-tax dollars, it becomes clear that home agricultural methods that take a significant chunk out of that figure can make a difference.

The key to making a garden work to your economic benefit is to approach mini-farming as a business. No, it is not a business in the sense of incorporation and taxes, unless some of its production is sold. But think of it as a business in that, by reducing your food expenditures, it can have the same net effect on finances as income from a small business. Like any small business, it could earn money or lose money depending on how it is managed.

Seeds and Seedlings

Garden centers are flooded every spring with gardeners buying seedlings. For hobbyist gardeners, this may work well because it allows a quick start with minimal planning. But for the mini-farmer who approaches gardening as a small business, it's a bad idea.

In my garden this year, I plan to grow 48 broccoli plants. Seedlings from the garden center would cost $18 if discounted, possibly more than $30. Even the most expensive organic broccoli seeds on the market cost less than a dollar for 48 seeds. Growing transplants at home drops their effective cost from $18 to $30 down to $1.

Adding the cost of soil and containers, the cost is still only about $2 for 48 broccoli seedlings. Considering that a mini-farm would require transplants for dozens of crops, from onion sets to tomatoes and lettuce, it quickly becomes apparent that growing from seed saves hundreds of dollars a year. (To learn more about choosing seeds or seedlings read, Should You Plant Seeds or Seedlings?)

The two basic types of seed/plant varieties available are hybrid and open-pollinated. Open-pollinated varieties produce seeds that duplicate the plants that produced them. Hybrid plant varieties produce seeds that are at best unreliable and sometimes sterile and therefore often unusable.

Although hybrids have the disadvantage of not producing good seed, they often have advantages that make them worthwhile, including aspects of "hybrid vigor," a poorly understood phenomenon in plants where a cross between two varieties can yield far more vigorous and productive offspring than either parent. Using hybridization, then, seed companies are able to deliver varieties that incorporate disease resistance into a particularly good-tasting variety. So why not just use hybrid seeds? Because there's no such thing as a free lunch. For plants that normally self-pollinate, such as peppers and tomatoes, there's no measurable increase in vigor in hybrids. The hybrids are just a marketing avenue — buying hybrids raises costs and forces you to buy seeds again next year.

Another reason to save seeds from open-pollinated plant varieties is that, if each year you save seeds from the best-performing plants, you will eventually create varieties with genetic characteristics that work best in your particular soil and climate. That's a degree of specialization money can't buy.

Of course, there are cases where hybrid seeds outperform open-pollinated varieties. Hybrid seeds that manifest pest- or disease-resistant traits can be a good choice when those pests or diseases cause ongoing problems. When using hybrid seeds eliminates the need for synthetic pesticides, they're a good choice. (Learn more in Heirloom Plants vs. New Plant Varieties.)

Compost

Because growing so many plants in such little space puts heavy demands on the soil, all intensive agriculture methodologies pay particular attention to maintaining soil fertility. Standard agribusiness practices would suggest buying commercial fertilizers from outside the farm. While there are other highly worthwhile reasons for avoiding the use of nonorganic fertilizers (including human health and environmental damage), economics alone make a good case for avoiding synthetic fertilizers. A mini-farm with a properly managed soil-fertility plan can drastically reduce the need to purchase fertilizer, thereby reducing one of the biggest costs associated with farming. A certain amount of fertilizer may always be required, especially at the beginning, but using organic fertilizers and creating compost can ultimately reduce fertilizer requirements to a bare minimum.

The practice of preserving soil fertility consists of growing crops specifically for compost value, growing crops to fix atmospheric nitrogen into the soil, and composting all crop residues possible (along with the specific compost crops) and practically anything else that isn't nailed down.

Calorie-Dense Plants: As already noted, vegetables provide only about 10 percent of the average American's calories. Because of this, a standard vegetable garden may supply excellent produce and rich vitamin content, but the economic value of the vegetables won't significantly reduce your food bill. The solution is to also grow crops that provide a higher proportion of caloric needs such as fruits, dried beans, grains, and root crops such as potatoes and onions.

Meat: Most Americans obtain at least a portion of their protein from eggs and meat. Agribusiness meats are often produced using practices and substances (such as growth hormones and antibiotics) that worry a lot of people. Certainly, factory-farmed meat is very high in the least healthy fats compared with free-range, grass-fed animals. The problem with meat, in an economic sense, is that each calorie of meat generally requires two to four calories of feed. This sounds, at first, like an inefficient use of resources, but it isn't as bad as it seems. Most livestock, including poultry, gets a substantial portion of its diet from foraging. Poultry will eat all of the ticks, fleas, spiders, beetles and grasshoppers that can be found, plus dispose of the farmer's table scraps. If meat is raised on premises, the mini-farmer just has to raise enough food to make up the difference between feed needs and what's obtained through scraps and foraging.

Fruit: A number of fruits can be grown in most parts of the country: apples, grapes, blackberries, pears and cherries, to name few. Dwarf fruit tree varieties often produce substantial amounts of fruit in only three years, and they take up comparatively little space. Grapes native to North America, such as the Concord grape, are hardy throughout the continental United States, and some varieties, such as muscadine grapes, grow prolifically in the South and offer unique health benefits. Strawberries are easy to grow and attractive to youngsters. Fruits can easily be preserved, and many can also be stored whole for a few months using root cellaring.

Market Crops: Especially if you adopt organic growing methods, you can get top wholesale-dollar for crops delivered to restaurants, food cooperatives, farmers markets and so forth. According to John Jeavon's research described in The Complete Biointensive Mini-Farm, a U.S. mini-farmer could expect to earn $2,079 in income from the space required to feed one person, in addition to actually feeding the person. Assuming a family of three and correcting for USDA reported rises in the value of food, that amounts to about $10,000 a year, using a six-month growing season.

Mel Bartholomew, in his 1985 book, Ca$h from Square Foot Gardening, estimated $5,000 a year income during a six-month growing season from a mere 1,500 square feet of properly managed garden. This equates to $8,064 in today's market. A mini-farm that sets aside only 2,100 square feet for market crops could gross an average of $11,289 per year. (It's worth noticing that two authorities arrived at very similar numbers for expected income from vegetable sales — about $5 a square foot.)

Raised Beds Farming

Raised bed farming refers to the agricultural technique of building freestanding crop beds above the existing level of soil. Sometimes raised beds are covered with plastic mulch to create a closed planting bed.

Raised beds may be rectangular in shape or have an irregular shape. They can be formal or informal in their design. A raised bed does not need to be very deep to be productive. Depending on the crop, farmers can build beds of only 8"-12" deep.

Two Types of Raised Bed Farm Design

A raised farm bed without support looks like a flat top mound about 6"-8" tall. It requires no additional materials other than the enriched soil. This type is more commonly used in large-scale and commercial farming.

Alternatively, a framework made of plastic, brick, stone, or untreated wood surrounds a supported raised bed. It looks like a planting box filled with soil. Home gardeners more commonly use supported raised beds.

Benefits of Raised Bed Farming

Raised bed farming techniques do not exist just to be different. They offer significant benefits to both small and large-scale farmers using them correctly. While the construction of raised beds differs when done in the small-scale versus the large-scale, the advantages remain the same.

Farmers trying to grow crops in rocky or poor soil must spend a lot of money installing drainage systems. The same goes for commercial farmers working in an area with too much rainfall or a high water table. Switching to a raised crop bed saves money since you do not need to invest in an expensive drainage system.

Commercial farmers who live in colder environments also benefit from a raised farm bed versus traditional techniques. A raised bed lifts the soil out of the hard ground into smaller piles with trenches between. In spring, this allows greater airflow to reach the soil so it warms up much faster.

It does not matter if you want to grow vegetables, herbs, flowers, or fruit. Any industrial agricultural operation will benefit from incorporating raised beds.

Two Types of Raised Bed Farm Design

A raised farm bed without support looks like a flat top mound about 6"-8" tall. It requires no additional materials other than the enriched soil. This type is more commonly used in large-scale and commercial farming.

Alternatively, a framework made of plastic, brick, stone, or untreated wood surrounds a supported raised bed. It looks like a planting box filled with soil. Home gardeners more commonly use supported raised beds.

Raised Bed Size Considerations

No doubt you've already been looking at dozens, if not hundreds, of images of other gardens. So, you know that bed sizes and shapes vary widely. I've seen just about everything too – even plants inserted directly into bags of garden soil (not something I recommend). Here are the guidelines I do recommend:

1. Height: 12-18" is ideal, however even as low as 6" can work and be productive. Most feeder roots are in the first 6", but the deeper the roots, the taller the shoots. Going higher than 18" can potentially cause more structural issues down the road – due to the weight and pressure of all that soil.

Think about what types of crops you want to grow (root vegetables which require more space, herbs which require less, etc.). Think too about the foundation on which you will be building. Will the surface allow the soil to erode out the bottom (go higher), or might it be impacted by the weight of the bed (don't go too high)?

2. Width: Four feet is perfect, but three feet can also work. Four feet allows more flexibility for spacing rows, but more importantly, not building beyond that width will allow you to easily reach the center from either side of the bed. It's important that you don't have to step into the bed to weed, plant, etc., as that will compact the soil and affect drainage and overall health.

3. Length: Whatever fits your needs. You could build 4'x4' squares. You could build 4'x20' rows. As long as you stick within a four-foot maximum width, your length is only limited by your space and budget.

4. Shape: As mentioned, you can build squares, rectangles, T's, circles, ovals, etc. As long as you can reach all areas of the bed from the edge (staying within that four-foot width), you're all set.

Which Materials are Safe for Containing your Beds?

Size and shape will likely also be dependent on your materials. More importantly, many of you expressed concerns and had questions about which materials are safe. Here's where things can really go sideways. There are so much conflicting information and surprisingly few studies on the various materials available for use.

Why do materials matter? First of all, the materials you use will be in close quarters to your food crop. In all likelihood, the roots and foliage will be regularly making contact with your material surface.

Secondly, the soil you place in your bed will need to remain fairly moist, and the exterior surfaces of your bed will be spending a lot of time in the hot sun. Most materials degrade when exposed to constant moisture and sunlight.

I used 16' lengths of 6"x6" untreated cedar at the GardenFarm but living in the heavily-populated Atlanta area offers me a better supply of wood materials than will be available for many of you.

Regardless, here are pros and cons to the materials you may be considering:

Raw Wood:

The best types of untreated wood are black walnut, cypress, cedar, redwood, oak, black locust, or osage orange. These are known for their rot-resistant properties and last for many years, even under moist conditions.

These woods can be difficult to find available for purchase in some areas. They are also expensive. Untreated pine is a less expensive untreated option, but it will also have a shorter lifespan.

Another consideration: Aside from pine, these woods are not as sustainable as other materials. Often, these woods are harvested from old-growth forest. If you choose to use one of these woods, check that it is coming from a sustainable source. Look for the Forest Stewardship Council (FSC) certification on any wood you buy. The FSC is an international organization that has developed standards for responsible forest management.

All types untreated wood will need to be replaced at some point. The lifespan of your wood will depend on wood type and your environment. If you live in an arid climate, untreated wood can last for several years. If you live in a hot and muggy area, untreated wood may only see you through a couple of years.

 Replacing your wood does not signify failure. The untreated wood is decomposing and even adding some nutrients to your garden bed in the process. It's more a matter of maintenance and realistically assessing what will work best for you and your family.

Wood Stains & Paint:

You may opt to extend the life of your untreated wood by staining or painting it. I recommend using a natural treatment like raw linseed oil or raw tung oil.

It's important to look for the raw form of these, as those not marked "raw" will likely include other chemicals. The chemicals are added to speed up the oil drying process, so by using the raw versions, allow for additional drying time.

Another thing to bear in mind is that linseed oil is a food source for mildew, so if mildew is a problem in your area, that may not be a good choice for you.

There haven't been many studies on the impact of using paints or stains for garden bed structure. Paint and stain ingredients vary, and overall, the impact is relatively unknown. But common sense should remind you that these all include chemicals of some nature, and those chemicals may impact your crop.

I recommend against painting the exterior only of your raised bed structure. The wood exposed to the moist soil will wick up moisture, but the exterior paint won't allow the wood to fully "breathe." So by painting the exterior only, you will be trapping the moisture inside and shortening the lifespan of your wood.

Treated Wood:

Treated wood has been infused with chemical elements to preserve the wood. CCA (Chromated Copper Arsenate) wood used to be the most commonly available. The primary concern with treated wood is that those infused elements leach out of the wood. The arsenic in CCA led manufacturers of CCA-treated wood to discontinue its availability for residential applications in 2003.

Although you may find older CCA-treated wood, today's retail options will more likely be ACQ (Ammoniacal Copper Quat) or MCA (Micronized Copper Azole). They have a higher concentration of copper but don't have the arsenic.

Leaching occurs at the highest levels under the following conditions:

- Smaller surfaces – i.e., the ends and – especially, the sawdust
- More recently treated (although CCA-treated wood is shown to retain uniformly-high levels of CCA)
- Moist conditions – i.e., after rain or in a muggy environment
- In unhealthy soil

So to put this into an interesting perspective, studies exploring the impact of treated wood when used for raised beds have shown that the greatest risk is actually in touching the exterior of the bed. When you (or especially, your kids) sit on or lean on treated wood, your skin or clothing is most likely to absorb the copper or arsenic leaching out of the wood to remain on the surface.

If you currently have beds made of the older CCA-treated wood, don't be alarmed. If you're using lots of compost, you should be fine, since plants don't even take up arsenic unless the soil is deficient in phosphorus. And that's likely not the case since phosphorus tends to be immobile and ongoing amendments of compost just add to the overall volume.

In other words, really healthy soil with lots of organic matter does not take up arsenic by plant roots. Yet the more acidic or alkaline your soil, the more likelihood of those elements being taken up by your plants. So, just another reason for getting a soil test to get your soil closer to a neutral pH (6.5-7.0 – also the ideal range for vegetable growth). Ditto for soil with a low amount of organic matter, so make sure your soil analysis tests for organic matter percentage as well.

As for the newer ACQ and MCA treated wood (which have higher copper levels), plants in your food garden won't be able to tolerate high levels of copper, and studies show that healthy soil also prevents uptake of copper.

Even if copper levels are high and being taken up, the plants will die before you ever have a chance to think about eating them. At any rate, that would be a good indicator of a potential problem – in which case you might want to think about having your soil tested for metal concentrations.

While we're on the subject, root vegetables are at greatest risk of being impacted by leaching, as most metals (when taken up) remain in plant roots. Studies further show that those root vegetables are impacted most on their surface. So by thoroughly washing all the impacted soil off and peeling the skin off your potatoes, beets, etc.; you will be eliminating potential contamination.

Your tomatoes and your eggplant could absorb copper or arsenic into their roots, but it is generally not shown to affect the fruit. Leafy greens are an exception and can take up arsenic in their leaves.

In short: Keep your soil near neutral and add lots of compost (more on both of these later), thoroughly wash off the soil and peel the skin from your root vegetables, and avoid contact with the exterior surface of the treated wood. As an extra precaution, grow leafy greens and root vegetables more toward the center of your bed (12" from the perimeter if possible), furthest from the treated wood.

A final note: When building treated wood beds, make your cuts somewhere that allows you to contain the sawdust. Wear a dust mask and gloves, and remove and dispose of the sawdust promptly. Don't add it to your compost.

Cinder or Concrete Blocks:

The truth is, these days the terms are used interchangeably. If your "cinder" blocks are decades old, they may actually be cinder blocks, but only concrete blocks have been in production for the past 50+ years.

What are your concrete blocks made from? That depends somewhat on your area, but there are consistencies. Virtually all concrete blocks are made of what's called Portland cement as well as aggregate, like sand or gravel.

One of the ingredients of Portland cement is fly ash (ranging from 15% to 25%). It's used to make concrete blocks lighter yet stronger. Fly ash is a fine powder byproduct of coal burning, so In other words, it's a petroleum byproduct.

And here's the real rub: fly ash contains various amounts of toxic metals; including arsenic, lead, and mercury. So, yes, those metals are in the concrete blocks that line your vegetable garden too.

While that might sound scary, the risk of those metals becoming available in the soil only happens if part of the concrete block is pulverized. Then, it's a matter of several factors that determine the potential risk to what you are growing.
First, the proximity of plant roots to the damaged area. Next, soils higher in organic matter are always beneficial but especially in this case, because they help chemically bind the metals – making them unavailable for absorption into the plant. Just as with CCA-treated wood, root crops and leafy greens are most susceptible when exposed to higher concentrations.
So, how much fly ash is absorbed by soil held within a concrete block structure? Well, if the block is intact, little to none. But not much research has been done on this specific subject.
If you have beds made from concrete blocks, just avoid anything that would cause them to break to the point that the dust from pulverized pieces can come in contact with plant roots.

And if you really want to "do something," seal the interior lining with a polymer paint (the most practical option), or line the interior side with PE plastic. It's up to you to decide if it's really worth the trouble.

If building raised beds over a concrete surface, the same risks and preventions would apply.

Composite Wood:

Composite wood is made of recycled materials and can last for years. Some composite material, when used in long side walls can buckle a bit. Here again, there hasn't been much research on the use of composite wood in proximity to edibles.

Are there any chemicals or elements released by the composite material? It appears to be a benign product for garden use, but there isn't much information out there to make a solid determination.

Railroad Ties:

Railroad ties are made with creosote, an oil distilled from coal tar. Creosote is used as a wood preservative for industrial use and is the black, oily stuff you see oozing from the sides of the ties.

The heft of railroad ties has made them a popular choice for raised beds and garden retaining walls. While there have been few studies on the impact of using them to contain edible plants, I'll take the advice provided directly from the EPA on creosote: "...Creosote is not approved to treat wood for residential use, including landscaping timbers or garden borders.... Creosote is a possible human carcinogen and has no registered residential use."

Galvanized Metal:
There's little scientific information available examining the effect of galvanized metal in the use of raised beds. What I can tell you is that the galvanization process typically involves dipping the metal in molten zinc or zinc-based coating.
While dangerous if consumed in large quantities, zinc is a micronutrient that plants and humans actually need in small quantities. If too much zinc were leached into the soil, it would probably reflect in dying plants, before it would ever pose a health risk.
Also, galvanized metal has been used to hold or transport water for humans and livestock for many years. All that to say: I can't guarantee there's not a negative health impact, but the risk is certainly low.

One thing you should bear in mind, however, is heat and drainage. Livestock troughs are a popular option, but it's critical to provide lots of drainage holes in the bottom of the trough. That moisture will need a place to escape, so you don't inadvertently drown your plant roots.

Whether you use metal sheeting or a trough, that metal will absorb and reflect heat from the sun – more than other materials. As a result, your soil will tend to dry out more quickly, and foliage in the line of that reflective power might suffer. The soil nearest to the sun-facing metal will also warm up more than the rest of the bed.

It might be wise to plant those tender vegetables – like lettuce – toward the center of the bed where soil temperature will remain most constant.

Tires:

Just don't. Don't do it. If you must do it, do it only for a season or two at most. Tires are a petroleum-based product. Their rubber degrades in the heat and moisture, and the chemicals incorporate into your soil. They may be convenient or look kitschy and fun. It may keep a tire out of the landfill, sure. But there are more drawbacks to using tires than there are benefits.

There's a reason that most landfills prohibit tires. If garbage shouldn't be subjected to decomposing tire rubber, neither should your family's food.

Pre-Made Kits:

It doesn't get much easier than one of the many raised bed kits available for purchase today. These can be used with composite wood and can be cut to varying lengths. Some can be expensive, and the material with which they are made can vary widely. I recently built raised beds on an episode of Growing a Greener World, so check that out.

Liners:

If you use any of the above materials with the potential of leaching, you might be inclined to line the bed with plastic. Yes, this will provide a barrier between the bed material and your soil. But don't lose sight of the plastic material itself.

There are so many plastics out there, and they are of widely-diverse safety grades. If you use plastic, look for food-grade polyethylene. This is considered one of the most food-safe plastics. Line only the outer perimeter of the bed – not the bottom surface. Don't block drainage with plastic.

With all of these products, I recommend you do your own research to feel comfortable in your choice. There are many reputable and not-so-reputable resources out there, so always be mindful of your information resource.

Soil Composition & Maintenance

Soil is one of the most important elements of an ecosystem, and it contains both biotic and abiotic factors. The composition of abiotic factors is particularly important as it can impact the biotic factors, such as what kinds of plants can grow in an ecosystem. Soil contains air, water, and minerals as well as plant and animal matter, both living and dead. These soil components fall into two categories. In the first category are biotic factors—all the living and once-living things in soil, such as plants and insects. The second category consists of abiotic factors, which include all nonliving things—for example, minerals, water, and air. The most common minerals found in soil that support plant growth are phosphorus, and potassium and also, nitrogen gas. Other, less common minerals include calcium, magnesium, and sulfur. The biotic and abiotic factors in the soil are what make up the soil's composition.

Soil composition is a mix of soil ingredients that varies from place to place. The Natural Resources Conservation Service (NRCS)—part of the U.S. Department of Agriculture— has compiled soil maps and data for 95 percent of the United States. The NRCS has found that each state has a "state soil" with a unique soil "recipe" that is specific to that state. These differing soils are the reason why there is such a wide variety of crops grown in the United States.

Consider the soils of three states: Hawai'i, Iowa, and Maine. Hawai'i's deep, well-drained state soil contains volcanic ash that makes it perfect for growing sugar cane, as well as ginger roots, papaya, and macadamia nuts. Iowa, which is in Midwest region of the United States, has a state soil that is good for farming because it is made up of a thick layer of organic matter from the decomposition of prairie grasses. Corn and soybeans are the primary crops grown in these soils. The state soil of Maine, located in the northeastern part of the country, is made from materials left behind after local glaciers melted. This soil is perfect for growing trees—specifically, red spruce and balsam fir. Many of the trees being grown today in Maine are harvested for timber or for making paper.

Soil scientists conduct various tests on soils to learn about their composition. Soil testing can identify the amounts of biotic and abiotic factors in the soil. The results of these tests can also reveal if the soil has too much of a specific mineral or if it needs more nutrients to support plants. Scientists also measure other factors, such as the amount of water in the soil and how it varies over time—for instance, is the soil unusually wet or dry? The tests can also identify contaminants and heavy metal in the soil and determine the soil's nitrogen content and pH level (acidity or alkalinity). All of these measurements can be used to determine the soil's health.

A typical soil is about 50% solids (45% mineral and 5% organic matter), and 50% voids (or pores) of which half is occupied by water and half by gas.[31] The percent soil mineral and organic content can be treated as a constant (in the short term), while the percent soil water and gas content is considered highly variable whereby a rise in one is simultaneously balanced by a reduction in the other. The pore space allows for the infiltration and movement of air and water, both of which are critical for life existing in soil. Compaction, a common problem with soils, reduces this space, preventing air and water from reaching plant roots and soil organisms.

Given sufficient time, an undifferentiated soil will evolve a soil profile which consists of two or more layers, referred to as soil horizons. These differ in one or more properties such as in their texture, structure, density, porosity, consistency, temperature, color, and reactivity. The horizons differ greatly in thickness and generally lack sharp boundaries; their development is dependent on the type of parent material, the processes that modify those parent materials, and the soil-forming factors that influence those processes. The biological influences on soil properties are strongest near the surface, while the geochemical influences on soil properties increase with depth. Mature soil profiles typically include three basic master horizons: A, B, and C. The solum normally includes the A and B horizons. The living component of the soil is largely confined to the solum, and is generally more prominent in the A horizon. It has been suggested that the pedon, a column of soil extending vertically from the surface to the underlying parent material and large enough to show the characteristics of all its horizons, could be subdivided in the humipedon (the living part, where most soil organisms are dwelling, corresponding to the humus form), the copedon (in intermediary position, where most weathering of minerals takes place) and the lithopedon (in contact with the subsoil).

The soil texture is determined by the relative proportions of the individual particles of sand, silt, and clay that make up the soil. The interaction of the individual mineral particles with organic matter, water, gases via biotic and abiotic processes causes those particles to flocculate (stick together) to form aggregates or peds. Where these aggregates can be identified, a soil can be said to be developed, and can be described further in terms of color, porosity, consistency, reaction (acidity), etc.

Water is a critical agent in soil development due to its involvement in the dissolution, precipitation, erosion, transport, and deposition of the materials of which a soil is composed. The mixture of water and dissolved or suspended materials that occupy the soil pore space is called the soil solution. Since soil water is never pure water, but contains hundreds of dissolved organic and mineral substances, it may be more accurately called the soil solution. Water is central to the dissolution, precipitation and leaching of minerals from the soil profile. Finally, water affects the type of vegetation that grows in a soil, which in turn affects the development of the soil, a complex feedback which is exemplified in the dynamics of banded vegetation patterns in semi-arid regions.

Soils supply plants with nutrients, most of which are held in place by particles of clay and organic matter (colloids). The nutrients may be adsorbed on clay mineral surfaces, bound within clay minerals (absorbed), or bound within organic compounds as part of the living organisms or dead soil organic matter. These bound nutrients interact with soil water to buffer the soil solution composition (attenuate changes in the soil solution) as soils wet up or dry out, as plants take up nutrients, as salts are leached, or as acids or alkalis are added.

Plant nutrient availability is affected by soil pH, which is a measure of the hydrogen ion activity in the soil solution. Soil pH is a function of many soil forming factors, and is generally lower (more acid) where weathering is more advanced.

Most plant nutrients, with the exception of nitrogen, originate from the minerals that make up the soil parent material. Some nitrogen originates from rain as dilute nitric acid and ammonia, but most of the nitrogen is available in soils as a result of nitrogen fixation by bacteria. Once in the soil-plant system, most nutrients are recycled through living organisms, plant and microbial residues (soil organic matter), mineral-bound forms, and the soil solution. Both living soil organisms (microbes, animals and plant roots) and soil organic matter are of critical importance to this recycling, and thereby to soil formation and soil fertility. Microbial soil enzymes may release nutrients from minerals or organic matter for use by plants and other microorganisms, sequester (incorporate) them into living cells, or cause their loss from the soil by volatilisation (loss to the atmosphere as gases) or leaching

Soil Maintenance

One of the keys to maintaining native and climate-appropriate plants in your sustainable landscape is the health of the soil. Plant roots utilize and store nutrients in the soil environment, so it is essential that the soil contains what the roots need!

Soil maintenance is the application of operations, practices, and treatments to protect soil and enhance its performance (such as soil fertility or soil mechanics). It includes soil conservation, soil amendment, and optimal soil health. In agriculture, some amount of soil management is needed both in nonorganic and organic types to prevent agricultural land from becoming poorly productive over decades. Organic farming in particular emphasizes optimal soil management, because it uses soil health as the exclusive or nearly exclusive source of its fertilization and pest control.

Soil maintenance is an important tool for addressing climate change by increasing soil carbon and as well as addressing other major environmental issues associated with modern industrial agriculture practices. Project Drawdown highlights three major soil management practices as actionable steps for climate change mitigation: improved nutrient management, conservation agriculture (including No-till agriculture), and use of regenerative agriculture.

Specific soil maintenance practices that affect soil health include:
• Controlling traffic on the soil surface helps to reduce soil compaction, which can reduce aeration and water infiltration.
• Planting cover crops that keep the soil anchored and covered in off-seasons so that the soil is not eroded by wind and rain.
• Crop rotations for row crops alternate high-residue crops with lower-residue crops to increase the amount of plant material left on the surface of the soil during the year to protect the soil from erosion.
• Nutrient management can help to improve the fertility of the soil and the amount of organic matter content, which improves soil structure and function.
• Tilling the soil, or tillage, is the breaking of soil, such as with a plough or harrow, to prepare the soil for new seeds. Tillage systems vary in intensity and disturbance. Conventional tillage is the most intense tillage system and disturbs the deepest level of soils. At least 30% of plant residue remains on the soil surface in conservation tillage. Reduced-tillage or no-till operations limit the amount of soil disturbance while cultivating a new crop, and help to maintain plant residues on the surface of the soil for erosion protection and water retention.
• Adding organic matter to the soil surface can increase carbon in the soil and the abundance and diversity of microbial organisms in the soil.
• Using fertilizers increases nutrients such as nitrogen, phosphorus, sulfur, and potassium in the soil. The use of fertilizers influences soil pH and often acidifies soils, with the exception of potassium fertilizer. Fertilizers can be organic or synthetic.

Compost

Compost is a mixture of ingredients used to fertilize and improve the soil. It is commonly prepared by decomposing plant and food waste and recycling organic materials. The resulting mixture is rich in plant nutrients and beneficial organisms, such as worms and fungal mycelium. Compost improves soil fertility in gardens, landscaping, horticulture, urban agriculture, and organic farming. The benefits of compost include providing nutrients to crops as fertilizer, acting as a soil conditioner, increasing the humus or humic acid contents of the soil, and introducing beneficial colonies of microbes that help to suppress pathogens in the soil. It also reduces expenses on commercial chemical fertilizers for recreational gardeners and commercial farmers alike.[1] Compost can also be used for land and stream reclamation, wetland construction, and landfill cover compost, crumbly mass of rotted organic matter made from decomposed plant material, used in gardening and agriculture. Compost is especially important in organic farming, where the use of synthetic fertilizers is not permitted. Compost improves soil structure, provides a wide range of nutrients for plants, and adds beneficial microbes to the soil. The maximum benefits of compost on soil structure (better aggregation, pore spacing, and water storage) and on crop yield usually occur after several years of use. Composts commonly contain about 2 percent nitrogen, 0.5–1 percent phosphorus, and about 2 percent potassium. Nitrogen fertilizers and manure may be added to speed decomposition. The nitrogen of compost becomes available slowly and in small amounts, which reduces leaching and extends availability over the whole growing season. Because of their fairly low nutrient content, composts are usually applied i large amounts. Compost can be prepared on a small scale for home gardens, usually in a simple pile of yard waste and kitchen scraps, though compost bins and barrels are also used. Aeration is important for proper decomposition, so piles are usually mixed every few days. When properly prepared, compost is free of obnoxious odours. A compost pile with the right ratio of carbon to nitrogen (30:1) and with adequate moisture will produce enough heat during decomposition to kill many pathogens and seeds, though it is advisable to avoid adding diseased plant matter and weeds that have gone to seed. Some municipalities collect household yard waste for large-scale composting, which reduces the amount of organic matter in landfills.

Vermicomposting is a method of composting that utilizes earthworms. Worms are kept in specialized bins and fed kitchen scraps and other plant matter. After several weeks the worms are removed, and their rich castings (manure) are collected for use as a soil amendment.

At the simplest level, composting requires gathering a mix of 'Greens' and 'Browns'. Greens are materials that are rich in nitrogen such as leaves, grass, and food scraps. Browns are more woody materials that are rich in carbon, such as stalks, paper, and wood chips. Materials are wetted to break them down into humus, a process that occurs for months.[citation needed] However, composting can also take place as a multi-step, closely monitored process with measured inputs of water, air, and carbon- and nitrogen-rich materials. The decomposition process is aided by shredding the plant matter, adding water, and ensuring proper aeration by regularly turning the mixture in a process that uses open piles or "windrows." Fungi, earthworms, and other detritivores further break up the organic material. Aerobic bacteria and fungi manage the chemical process by converting the inputs into heat, carbon dioxide, and ammonium. Composting is an important part of waste management since food and other compostable materials make up about 20% of waste in landfills and these materials take longer to biodegrade in the landfill. Composting offers an environmentally superior alternative to using organic material for landfill because composting reduces methane production, and provides economic and environmental co-benefits

Plant Nutrient In Depth

To understand the importance of plants in structuring the vertical distributions of soil nutrients, we explored nutrient distributions in the top meter of soil for more than 10,000 profiles across a range of ecological conditions. Hypothesizing that vertical nutrient distributions are dominated by plant cycling relative to leaching, weathering dissolution, and atmospheric deposition, we examined three predictions: (1) that the nutrients that are most limiting for plants would have the shallowest average distributions across ecosystems, (2) that the vertical distribution of a limiting nutrient would be shallower as the nutrient became more scarce, and (3) that along a gradient of soil types with increasing weathering-leaching intensity, limiting nutrients would be relatively more abundant due to preferential cycling by plants. Globally, the ranking of vertical distributions among nutrients was shallowest to deepest in the following order: P > K > Ca > Mg > Na = Cl = SO4. Nutrients strongly cycled by plants, such as P and K, were more concentrated in the topsoil (upper 20 cm) than were nutrients usually less limiting for plants such as Na and Cl. The topsoil concentrations of all nutrients except Na were higher in the soil profiles where the elements were more scarce. Along a gradient of weathering-leaching intensity (Aridisols to Mollisols to Ultisols), total base saturation decreased but the relative contribution of exchangeable K+ to base saturation increased. These patterns are difficult to explain without considering the upward transport of nutrients by plant uptake and cycling. Shallower distributions for P and K, together with negative associations between abundance and topsoil accumulation, support the idea that plant cycling exerts a dominant control on the vertical distribution of the most limiting elements for plants (those required in high amounts in relation to soil supply). Plant characteristics like tissue stoichiometry, biomass cycling rates, above- and belowground allocation, root distributions, and maximum rooting depth may all play an important role in shaping nutrient profiles. Such vertical patterns yield insight into the patterns and processes of nutrient cycling through time.

Plant Nutrients In The Soil

Soil is a major source of nutrients needed by plants for growth. The three main nutrients are nitrogen (N), phosphorus (P) and potassium (K). Together they make up the trio known as NPK. Other important nutrients are calcium, magnesium and sulfur. Plants also need small quantities of iron, manganese, zinc, copper, boron and molybdenum, known as trace elements because only traces are needed by the plant. The role these nutrients play in plant growth is complex, and this document provides only a brief outline.

Major Elements
Nitrogen (N)

Nitrogen is a key element in plant growth. It is found in all plant cells, in plant proteins and hormones, and in chlorophyll.

Atmospheric nitrogen is a source of soil nitrogen. Some plants such as legumes fix atmospheric nitrogen in their roots; otherwise fertiliser factories use nitrogen from the air to make ammonium sulfate, ammonium nitrate and urea. When applied to soil, nitrogen is converted to mineral form, nitrate, so that plants can take it up.

Soils high in organic matter such as chocolate soils are generally higher in nitrogen than podzolic soils. Nitrate is easily leached out of soil by heavy rain, resulting in soil acidification. You need to apply nitrogen in small amounts often so that plants use all of it, or in organic form such as composted manure, so that leaching is reduced.

Phosphorus (P)

Phosphorus helps transfer energy from sunlight to plants, stimulates early root and plant growth, and hastens maturity.

Very few Australian soils have enough phosphorus for sustained crop and pasture production and the North Coast is no exception. The most common phosphorus source on the North Coast is superphosphate, made from rock phosphate and sulfuric acid. All manures contain phosphorus; manure from grain-fed animals is a particularly rich source.

Potassium (K)

Potassium increases vigour and disease resistance of plants, helps form and move starches, sugars and oils in plants, and can improve fruit quality.

Potassium is low or deficient on many of the sandier soils of the North Coast. Also, heavy potassium removal can occur on soils used for intensive grazing and intensive horticultural crops (such as bananas and custard apples).

Muriate of potash and sulfate of potash are the most common sources of potassium.

Calcium (Ca)

Calcium is essential for root health, growth of new roots and root hairs, and the development of leaves. It is generally in short supply in the North Coast's acid soils.

Lime, gypsum, dolomite and superphosphate (a mixture of calcium phosphate and calcium sulfate) all supply calcium. Lime is the cheapest and most suitable option for the North Coast; dolomite is useful for magnesium and calcium deficiencies, but if used over a long period will unbalance the calcium/magnesium ratio. Superphosphate is useful where calcium and phosphorus are needed.

Magnesium (Mg)

Magnesium is a key component of chlorophyll, the green colouring material of plants, and is vital for photosynthesis (the conversion of the sun's energy to food for the plant). Deficiencies occur mainly on sandy acid soils in high rainfall areas, especially if used for intensive horticulture or dairying. Heavy applications of potassium in fertilisers can also produce magnesium deficiency, so banana growers need to watch magnesium levels because bananas are big potassium users.

Magnesium deficiency can be overcome with dolomite (a mixed magnesium-calcium carbonate), magnesite (magnesium oxide) or epsom salts (magnesium sulfate).

Sulfur (S)
Sulfur is a constituent of amino acids in plant proteins and is involved in energy-producing processes in plants. It is responsible for many flavour and odour compounds in plants such as the aroma of onions and cabbage.

Sulfur deficiency is not a problem in soils high in organic matter, but it leaches easily. On the North Coast seaspray is a major source of atmospheric sulfur. Superphosphate, gypsum, elemental sulfur and sulfate of ammonia are the main fertiliser sources.

Time & Yield

Time plays an important role in Agriculture. Everything can weight but not agriculture By Pt. Nehru. In India, the cultivation of crops is monsoon based that's why time having very significant role. The suitable climatic conditions for a particular crop are associated with the on set or off set of monsoon. And for the all cultural practices, favourable weather conditions are also attached with climate of a particular region. The occurrence of insect-pest and diseases is affected by moisture regime and temperature of the particular area for growing of the crops. Duration of crop growth cycles are above all, related to temperature. An increase in temperature will speed up development. In the case of an annual crop, the duration between sowing and harvesting will shorten (for example, the duration in order to harvest corn could shorten between one and four weeks). The shortening of such a cycle could have an adverse effect on productivity because senescence would occur sooner. When wheat is sown in the month of December, there is a drastic reduction in yield by each day causes reduction of 41.60 kg in north western plains and 56 kg per hectare in north-eastern plains of the country; Source: Annual report of All India Co-ordinated Wheat Improvement Project 1972-73, 1973-74. And when we apply isoproturon after 35 days of sowing of wheat, the Phalaris minor will not be killed by it. In spite of the development of the best production technology, late sowing usually results in a poor stand as well as the inadequate vegetative growth of the crop. The reproductive development in late sown wheat is also poor on account of the quick rise of ambient temperature towards the maturity stage.

Crop yield

In agriculture, the yield is a measurement of the amount of a crop grown, or product such as wool, meat or milk produced, per unit area of land. The seed ratio is another way of calculating yields.

Innovations, such as the use of fertilizer, the creation of better farming tools, new methods of farming and improved crop varieties, have improved yields. The higher the yield and more intensive use of the farmland, the higher the productivity and profitability of a farm; this increases the well-being of farming families. Surplus crops beyond the needs of subsistence agriculture can be sold or bartered. The more grain or fodder a farmer can produce, the more draft animals such as horses and oxen could be supported and harnessed for labour and production of manure. Increased crop yields also means fewer hands are needed on farm, freeing them for industry and commerce. This, in turn, led to the formation and growth of cities, which then translated into an increased demand for foodstuffs or other agricultural products.

Measurement

The units by which the yield of a crop is usually measured today are kilograms per hectare or bushels per acre.

Long-term cereal yields in the United Kingdom were some 500 kg/ha in Medieval times, jumping to 2000 kg/ha in the Industrial Revolution, and jumping again to 8000 kg/ha in the Green Revolution. Each technological advance increasing the crop yield also reduces the society's ecological footprint.
Yields are related to agricultural productivity, but are not synonymous. Agricultural productivity is measured in money produced per unit of land, but yields are measured in the weight of the crop produced per unit of land. A farmer can invest a large amount of money to increase his yields by a few percent, for example with an extremely expensive fertilizer, but if that cost is so high that it does not produce a comparative return on investment, his profits decline, and the higher yield can mean a lower agricultural productivity in this case. A yield is a 'partial measure of productivity', because it may fail to accurately measure the actual productivity of the farming operation by not including the totality of the inputs.

Seed ratio
The seed ratio is the ratio between the investment in seed versus the yield. For example, if three grains are harvested for each grain seeded, the resulting seed ratio is 1:3,[citation needed] which is considered by some agronomists as the minimum required to sustain human life.[3] One of the three seeds must be set aside for the next planting season, the remaining two either consumed by the grower, or for livestock feed. In parts of Europe the seed ratio during the 9th century was merely 1:2.5, in the Low Countries it improved to 1:14 with the introduction of the three-field system of crop rotation around the 14th century

Watering & Irrigation

Irrigation is the agricultural process of applying controlled amounts of water to land to assist in the production of crops,[1] as well as to grow landscape plants and lawns, where it may be known as watering. Agriculture that does not use irrigation but instead relies only on direct rainfall is referred to as rain-fed. Irrigation has been a central feature of agriculture for over 5,000 years and has been developed independently by many cultures across the globe.

Irrigation helps to grow agricultural crops, maintain landscapes, and revegetate disturbed soils in dry areas and during periods of less than average rainfall. Irrigation also has other uses in crop production, including frost protection,[2] suppressing weed growth in grain fields[3] and preventing soil consolidation.[4] Irrigation systems are also used for cooling livestock, dust suppression, disposal of sewage, and in mining. Irrigation is often studied together with drainage, which is the removal of surface and sub-surface water from a given location.

Efficient watering practices are one of the best ways to conserve water both during and between droughts.

⍰ Bubblers

Bubbler irrigation is a localized, low pressure, solid permanent installation system used for watering plants and shrubs. Each plant has a round or square basin which is flooded with water during irrigation. The system uses bubblers to produce low volumes of water that can efficiently infiltrate into the soil and wet the root zone.

⍰ Drip Irrigation?

Drip irrigation is a method of watering that applies water directly to the root area of plants through a system of tubing. It is ideal for drought conditions because it is both efficient and economical. The lines can become clogged so many professionals prefer to use bubblers.

- Drip irrigation is usually more than 90% efficient at allowing plants to use the water given, because water is applied slowly and directly to the root zone. In contrast, sprinklers are at best 65-75% efficient.
- Drip irrigation reduces evaporation and runoff thus reducing the total amount of water used.
- Saves time, money, and water because the system is so efficient
- Can be easily installed on diverse types and shapes of terrain
- Keeps moisture at an optimal range, increasing plant health and productivity
- Reduces weed growth by keeping other areas dry

Additional Ways to Save

The optimal time for watering your lawn is the early morning (around 5 a.m.) due to: the lack of evaporation that takes place and "morning dew" that adds to the moisture. It also allows the water to soak deeper while allowing the landscaping to dry in the day time to prevent harmful lawn diseases from too much moisture.

• Check sprinklers for leaks and make sure they are spraying on plants and not impermeable surfaces, like sidewalks or driveways, or install more efficient watering systems like drip or bubblers.

• Don't use sprinkler systems on windy days

• Plant drought tolerant landscaping. Save 30-60 gallons every time you water 1,000 square ft.

• Put mulch around existing plants to act as an insulator. You will have to water less frequently; saving 20-30 gallons per 1,000 square ft. Organic mulch will also improve soil and prevent weeds

• Know how much water your plants actually need. Many species need little to no water every day

Modern irrigation system planning and construction

Water supply

The first consideration in planning an irrigation project is developing a water supply. Water supplies may be classified as surface or subsurface. Though both surface and subsurface water come from precipitation such as rain or snow, it is far more difficult to determine the origin of subsurface water.

In planning a surface water supply, extensive studies must be made of the flow in the stream or river that will be used. If the streamflow has been measured regularly over a long period, including times of drought and flood, the studies are greatly simplified. From streamflow data, determinations can be made of the minimum, maximum, average daily, and average monthly flows; the size of dams, spillways, and downstream channel; and the seasonal and carry-over storage needed. If adequate streamflow data are not available, the streamflow may be estimated from rain and snow data, or from flow data from nearby streams that have similar climatic and physiographic conditions.

The quality, as well as the quantity, of surface water is a factor. The two most important considerations are the amount of silt carried and the kind and amount of salts dissolved in the water. If the silt content is high, sediment will be deposited in the reservoir, increasing maintenance costs and decreasing useful life periods. If the salt concentration is high, it may damage crops or accumulate in the soil and eventually render it unproductive.

Subsurface sources of water must be as carefully investigated as surface sources. In general, less is known about subsurface supplies of water than about surface supplies, so, therefore, subsurface supplies are harder to investigate. Engineers planning a project need to know the extent of the basic geological source of water (the aquifer), as well as the amount the water level is lowered by pumping and the rate of recharge of the aquifer. Often the only way for the engineer to obtain these data reliably is to drill test wells and make on-site measurements. Ideally, a project is planned so as not to use more subsurface water than is recharged. Otherwise, the water is said to be "mined," meaning that it is being used up as a natural resource and its use is considered unsustainable.

Two sources of water not often thought of by the layperson are dew and sewage or wastewater. In certain parts of the world, Israel and part of Australia, for example, where atmospheric conditions are right, sufficient dew may be trapped at night to provide water for irrigation. Elsewhere the supply of wastewater from some industries and municipalities is sufficient to irrigate relatively small acreages. Recently, due to greater emphasis on purer water in streams, there has been increased interest in this latter practice.

In some countries (Egypt for example) sewage is a valuable source of water. In others, such as the United States, irrigation is looked upon as a means of disposing of sewer water as a final step in the wastewater treatment process. Unless the water contains unusual chemical salts, such as sodium, it is generally of satisfactory quality for agricultural irrigation. Where the practice is used primarily as a means of disposal, large areas are involved and the choice of crop is critical. Usually only grass or trees can withstand the year-round applications.

Before a water supply can be assured, the right to it must be determined. Countries and states have widely varying laws and customs that determine ownership of water. If the development of a water supply is for a single purpose, then the determination of ownership may be relatively simple; but if the development is multipurpose, as most modern developments are, ownership may be difficult to determine, and agreements must be worked out among countries, states, municipalities, and private owners.

The area that can be irrigated by a water supply depends on the weather, the type of crop grown, and the soil. Numerous methods have been developed to evaluate these factors and predict average annual volume of rainfall needed. Some representative annual amounts of rainfall needed for cropland in the western United States are 305 to 760 mm (12 to 30 inches) for cereal grains and 610 to 1,525 mm (24 to 60 inches) for forage. In the Near East, cotton needs about 915 mm (36 inches), whereas rice may require two to three times that amount. In humid regions of the United States, where irrigation supplements rainfall, grain crops may require 150 to 230 mm (6 to 9 inches) of water. In addition to satisfying the needs of the crop, allowances must be made for water lost directly to evaporation and during transport to the fields.

Transport systems

The type of transport system used for an irrigation project is often determined by the source of the water supply. If a surface water supply is used, a large canal or pipeline system is usually required to carry the water to the farms because the reservoir is likely to be distant from the point of use. If subsurface water drawn from wells is used, a much smaller transport system is needed, though canals or pipelines may be used. The transport system will depend as far as possible on gravity flow, supplemented if necessary by pumping. From the mains, water flows into branches, or laterals, and finally to distributors that serve groups of farms. Many auxiliary structures are required, including weirs (flow-diversion dams), sluices, and other types of dams. Canals are normally lined with concrete to prevent seepage losses, control weed growth, eliminate erosion hazards, and reduce maintenance. The most common type of concrete canal construction is by slip forming. In this type of construction, the canal is excavated to the exact cross section desired and the concrete placed on the earth sides and bottom.

Pipelines may be constructed of many types of material. The larger lines are usually concrete whereas laterals may be concrete, cement–asbestos, rigid plastic, aluminum, or steel. Although pipelines are more costly than open conduits, they do not require land after construction, suffer little evaporation loss, and are not troubled by algae growth.

Water application

After water reaches the farm it may be applied by surface, subsurface, or sprinkler-irrigation methods. Surface irrigation is normally used only where the land has been graded so that uniform slopes exist. Land grading is not necessary for other methods. Each method includes several variations, only the more common of which are considered here.

Crop Proportion & Sizing

This study presents a crop planting and type proportion (CPTP) method for crop acreage estimation of complex and diverse agricultural landscapes. CPTP has three major components: (1) Crop planting proportion (CPP), estimated with wide-swath satellite remote sensing data to completely cover the monitoring area by segmenting cropped and non-cropped areas through unsupervised classification. (2) Crop type proportion (CTP), estimated by transect sampling and a special GPS-Video-GIS instrument (GVG) and a visual interpretation of crop type proportion in collected pictures for different strata. (3) Multiplication of CPP and CTP with arable land area at the strata level, summed to the province and national level. Validation has been done with in situ data for different agricultural landscapes over China. Both CPP estimation with remote sensing data and CTP estimation through ground survey have a high accuracy with average relative error (RE) and root mean square error (RMSE) equal to 1.42% and 1.67% for CPP and to 2.63% and 2.25% for CTP. The RE for crop acreage estimation equals to 4.09%. The CPTP method thus has a high accuracy, yields timely information at low costs, and is robust and provides objective results. The study concludes that the CPTP method can be used for large area crop acreage estimation of complex agriculture landscapes.

Numerous sources provide evidence of trends and patterns in average farm size and farmland distribution worldwide, but they often lack documentation, are in some cases out of date, and do not provide comprehensive global and comparative regional estimates. This article uses agricultural census data (provided at the country level in Web Appendix) to show that there are more than 570 million farms worldwide, most of which are small and family-operated. It shows that small farms (less than 2 ha) operate about 12% and family farms about 75% of the world's agricultural land. It shows that average farm size decreased in most low- and lower-middle-income countries for which data are available from 1960 to 2000, whereas average farm sizes increased from 1960 to 2000 in some upper-middle-income countries and in nearly all high-income countries for which we have information.

Such estimates help inform agricultural development strategies, although the estimates are limited by the data available. Continued efforts to enhance the collection and dissemination of up-to date, comprehensive, and more standardized agricultural census data, including at the farm and national level, are essential to having a more representative picture of the number of farms, small farms, and family farms as well as changes in farm size and farmland distribution worldwide.

Pest & Disease Control

Beginnings of pest control

Wherever agriculture has been practiced, pests have attacked, destroying part or even all of the crop. In modern usage, the term pest includes animals (mostly insects), fungi, plants, bacteria, and viruses. Human efforts to control pests have a long history. Even in Neolithic times (about 7000 bp), farmers practiced a crude form of biological pest control involving the more or less unconscious selection of seed from resistant plants. Severe locust attacks in the Nile Valley during the 13th century bp are dramatically described in the Bible, and, in his Natural History, the Roman author Pliny the Elder describes picking insects from plants by hand and spraying. The scientific study of pests was not undertaken until the 17th and 18th centuries. The first successful large-scale conquest of a pest by chemical means was the control of the vine powdery mildew (Unciluna necator) in Europe in the 1840s. The disease, brought from the Americas, was controlled first by spraying with lime sulfur and, subsequently, by sulfur dusting.

Another serious epidemic was the potato blight that caused famine in Ireland in 1845 and some subsequent years and severe losses in many other parts of Europe and the United States. Insects and fungi from Europe became serious pests in the United States, too. Among these were the European corn borer, the gypsy moth, and the chestnut blight, which practically annihilated that tree.

The first book to deal with pests in a scientific way was John Curtis's Farm Insects, published in 1860. Though farmers were well aware that insects caused losses, Curtis was the first writer to call attention to their significant economic impact. The successful battle for control of the Colorado potato beetle (Leptinotarsa decemlineata) of the western United States also occurred in the 19th century. When miners and pioneers brought the potato into the Colorado region, the beetle fell upon this crop and became a severe pest, spreading steadily eastward and devastating crops, until it reached the Atlantic. It crossed the ocean and eventually established itself in Europe. But an American entomologist in 1877 found a practical control method consisting of spraying with water-insoluble chemicals such as London Purple, paris green, and calcium and lead arsenates.

Other pesticides that were developed soon thereafter included nicotine, pyrethrum, derris, quassia, and tar oils, first used, albeit unsuccessfully, in 1870 against the winter eggs of the Phylloxera plant louse. The Bordeaux mixture fungicide (copper sulfate and lime), discovered accidentally in 1882, was used successfully against vine downy mildew; this compound is still employed to combat it and potato blight. Since many insecticides available in the 19th century were comparatively weak, other pest-control methods were used as well. A species of ladybird beetle, Rodolia cardinalis, was imported from Australia to California, where it controlled the cottony-cushion scale then threatening to destroy the citrus industry. A moth introduced into Australia destroyed the prickly pear, which had made millions of acres of pasture useless for grazing. In the 1880s the European grapevine was saved from destruction by grape phylloxera through the simple expedient of grafting it onto certain resistant American rootstocks.

This period of the late 19th and early 20th centuries was thus characterized by increasing awareness of the possibilities of avoiding losses from pests, by the rise of firms specializing in pesticide manufacture, and by development of better application machinery.

Integrated control

Some research into biological methods was undertaken by governments, and in many countries plant breeders began to develop and patent new pest-resistant plant varieties. One method of biological control involved the breeding and release of males sterilized by means of gamma rays. Though sexually potent, such insects have inactive sperm. Released among the wild population, they mate with the females, who either lay sterile eggs or none at all. The method was used with considerable success against the screwworm, a pest of cattle, in Texas. A second method of biological control employed lethal genes. It is sometimes possible to introduce a lethal or weakening gene into a pest population, leading to the breeding of intersex (effectively neuter) moths or a predominance of males. Various studies have also been made on the chemical identification of substances attracting pests to the opposite sex or to food. With such substances traps can be devised that attract only a specific pest species. Finally, certain chemicals have been fed to insects to sterilize them. Used in connection with a food lure, these can lead to the elimination of a pest from an area. Chemicals tested so far, however, have been considered too dangerous to humans and other mammals for any general use. Some countries (notably the United States, Sweden, and the United Kingdom) have partly or wholly banned the use of DDT because of its persistence and accumulation in human body fat and its effect on wildlife. New pesticides of lesser human toxicity have been found, one of the most used being mercaptosuccinate, trade named Malathion. A more recent important discovery was the systemic fungicide, absorbed by the plant and transmitted throughout it, making it resistant to certain diseases.

The majority of pesticides are sprayed on crops as solutions or suspensions in water. Spraying machinery has developed from the small hand syringes and "garden engines" of the 18th century to the very powerful "autoblast machines" of the 1950s that were capable of applying up to some 400 gallons per acre (4,000 litres per hectare). Though spraying suspended or dissolved pesticide was effective, it involved moving a great quantity of inert material for only a relatively small amount of active ingredient. Low-volume spraying was invented about 1950, particularly for the application of herbicides, in which 10 or 20 gallons of water, transformed into fine drops, would carry the pesticide. Ultralow-volume spraying has also been introduced; four ounces (about 110 grams) of the active ingredient itself (usually Malathion) are applied to an acre from aircraft. The spray as applied is invisible to the naked eye.

Barriers
Barriers are physical structures put in place to prevent a pest from reaching a plant. They keep pests away from a plant but do not kill them. Here are some examples that you can adapt, depending on the resources available to you:

Crawling insects
Cut the top off a transparent plastic bottle and place it firmly into the ground, over a young plant. This stops pests such as slugs from reaching the plant.
Using an old plastic bottle to protect a young plant

Climbing insects
To help protect trees from attack by insects, grease bands can be used.
Wrap a piece of plastic or a long leaf around the trunk of the tree. Spread any kind of thick grease on top of this. Fold over the top of the foil or plastic to form an overhang to protect the grease from being washed away by rain.
Check the grease every week to ensure that the grease is intact. This prevents crawling insects such as ants, fruit fly larvae, slugs, snails, beetles or caterpillars from damaging trees, especially fruit trees, or grain stores.

Termites
Digging a 70-100cm trench around buildings and nurseries can prevent attack from subterranean species of termites. This is a good method of control however it is hard work. Alternatively, barriers can be built. These should be partially above and below ground and should be made from material that is impenetrable to termites such as basalt, sand or crushed volcanic cinders.
Particle size of the material is critical, they should not be too large for the termites to carry away, and not so small that termites can pack the particles to create a continuous passage through which they can move.

Chemical control

Chemical pesticides are often used to control diseases, pests or weeds. Chemical control is based on substances that are toxic (poisonous) to the pests involved. When chemical pesticides are applied to protect plants from pests, diseases or overgrowth by weeds, we speak of plant protection products. It is of course important that the plant that needs protection does not itself suffer from the toxic effects of the protection products.

Efforts to protect crops started centuries ago. The Chinese, in around 1200 BC, used lime and wood ash to destroy parasites. The Romans used sulphur and bitumen, a substance derived from crude oil. Substances such as nicotine from tobacco were used from the 16th century on and later copper, lead and mercury as well. After the Second World War the use of true chemical pesticides started and nowadays there are hundreds of chemical pesticides available for use in agriculture and horticulture.
Pesticides are grouped into five main categories depending on the purpose they are usually applied for. The first group are the fungicides, which are act against fungi. Then there are herbicides which are used against weeds. Herbicides are taken up by the leaves or the roots of the weed, causing it to die. Insecticides that, as the name suggests, destroy harmful insects, and then there are acaricides which protect plants from mites. Finally there are nematicides to control nematodes that attack the plants.

The advantages and disadvantages of chemical pesticides

The use of chemical pesticides is widespread due to their relatively low cost, the ease
with which they can be applied and their effectiveness, availability and stability.
Chemical pesticides are generally fast-acting, which limits the damage done to crops.
Chemical pesticides have some major drawbacks, but they are still widely sold and used.
We will discuss four of the disadvantages of chemical pesticides here. First, chemical
pesticides are often not just toxic to the organisms for which they were intended, but also
to other organisms. Chemical pesticides can be subdivided into two groups: non-selective
and selective pesticides. The non-selective products are the most harmful, because they
kill all kinds of organisms, including harmless and useful species. For example, there are
herbicides that kill both broad-leaf weeds and grasses. This means they are non-selective
since they kill nearly all vegetation.
Selective pesticides have a more limited range. They only get rid of the target pest,
disease or weed and other organisms are not affected. An example is a weed killer that
only works on broadleaf weeds. This could be used on lawns, for example, since it does
not kill grass. These days, a combination of several products is usually required to control
several pests because almost all permitted products are selective and thus only control a
limited range of pests.
Another disadvantage of chemical pesticides is resistance. Pesticides are often effective
for only a (short) period on a particular organism. Organisms can become immune to a
substance, so they no longer have an effect. These organisms mutate and become
resistant. This means that other pesticides need to be used to control them.
A third drawback is accumulation. If sprayed plants are eaten by an organism, and that
organism is then eaten by another, the chemicals are can be passed up the food chain.
Animals at the top of the food chain, usually predators or humans, have a greater chance
of toxicity due to the build-up of pesticides in their system. Gradually, however, this
effect is becoming less relevant because pesticides are now required to break down more
quickly so that they cannot accumulate. If they do not, they are not permitted for sale.

Biological control
Biological controls consist of three different parts;
1. Macrobials
2. Microbials
3. Biochemicals
All three of these will be explained in brief.
Biological control using natural predators or parasites (macrobials)

Biological control is no fad. In China in the fourth century B.C., ants were used as the natural enemy of pest insects, and in South China today ants are still used to control pests in orchards and food stores. The usefulness of parasites was discovered much later. Most parasites are insects, such as parasitic wasps (Encarsia formosa), which during the egg, larva and pupa stages live in or on a host. The complicated life cycle of these insects was first described in the early 18th century by Antonie van Leeuwenhoek. However, it would be many years before their potential use in pest control was discovered. In 1800, Erasmus Darwin, the father of Charles Darwin, wrote an essay on the useful role that parasites and predators may play in combating pests and diseases.

Biological control assumes that natural predators or parasites are able to suppress pests. Initially, therefore, natural enemies were imported to control the pests. These natural predators were released in small numbers, but once they became established they were effective in the long term. This method is also called inoculation. When the natural predator is introduced periodically, it is known as inundation.

There are two groups of beneficial macrobial organisms: predators and parasites. Parasites are organisms that live at the expense of another organism, such as the larvae of parasitic wasps, which live in the larva of whitefly and eat them from the inside.

Predators are organisms that simply prey on other organisms for food, such as ladybirds, which eat aphids.

Some examples of commonly used macrobials are; Phytoseiulus persimilis against the red spider mite, Encarsia formosa against whitefly and Neioseiulus cucumeris against thrips.

Biological control using micro-organisms (microbials)

Several beneficial micro-organisms can also be used to improve plant health and control pests and diseases. Bacteria, fungi and other micro-organisms can have these effects because they compete for nutrients or space, they produce antibiotics or they simply eat other harmful micro-organisms.

Microbials can also be used preventively because they can make the plants healthier and stronger. When this happens, plants are not attacked by pests or diseases or are affected less by them. This kind of pest control is not visible.

Some examples of commonly used microbials are; Trichoderma and Bacillus subtilis. This is a coloured scanning electron micrograph (SEM) of Bacillus subtilis; a commonly used microbial. Microbials – micro-organisms that can be used for biological control – can make plants healthy and control pests and diseases. They can also be used preventively.

Biological control using resources of natural origin and pheromones (biochemicals)

In addition to macro-organisms and micro-organisms, there are also resources of natural origin and pheromones which can be used to control pests and diseases. This category is very wide, including plant extracts, vitamins and plant hormones. These also work preventively to make plants strong and healthy. The pheromones are used to lure the pest (insects) into a trap. Sex pheromones and aggregating pheromones are the most commonly used types.

The advantages and disadvantages of biological control
Biological control, just like chemical control, has advantages and disadvantages. We will
mention three major advantages here, as well as several disadvantages. The first
advantage is that the natural enemy can become established and this will produce long-
term results. The risk of resistance is also much lower since pests cannot build up
resistance to being eaten. Natural pest control is very targeted and therefore an effective
way to control particular pests.

The disadvantages of biological control are that natural enemies may move away. In
greenhouses this problem can be managed, but not in open fields. Spreading over a larger
plot also takes time. In the second place, pests are never destroyed completely because
the natural enemy needs to stay alive and they will therefore never destroy the entire
population. Finally, it is not possible to use them before the pest has occurred and this
means that some damage will be done to crops.
Some biological applications are not completely harmless either. Although these are
natural products, other organisms than those targeted may be harmed. A natural enemy
may also damage the crop, especially when large numbers are needed to control a pest.
The effect of natural enemies is also less pronounced than chemical control. So if the
biological method does not work, a higher dosage of chemical pesticides is required,
because the pest already is widely spread.
Finally there are no natural methods for the control of viruses others than removing the
affected plants.
Just like chemical control, biological control is constantly under development because
new pest organisms (insects, fungi, bacteria) appear and organisms mutate. Products that
provide biological control through chemicals of natural origin are classified as plant
protection products, just as pesticides are, and so they also have to meet strict
requirements. This category of 'plant protection products' can also be rather expensive as
a result.

Seed Starting

Seriously, when done right, seed starting can be an amazing way to increase the number of crops you can grow and your growing season, but not all seeds should be started indoors and there are many mistakes that can easily be avoided if you simply know ahead of time.

Seed Starting vs. Direct Sowing
What is the difference between starting seeds indoors and direct sowing seeds? In short, seed starting inside is where you germinate your seeds inside in a controlled environment, then transplant the young plants into the garden. Direct sowing seeds is done outside, where you plant the seeds directly into the soil where the plant will grow and live its entire life.

Why Start Seeds Indoors?
When you start seeds indoors you have more variety. If you're buying plant starts from a nursery, you're left to choose from the varieties they offer. But when you start from seed, you have so many more options.

You also are able to have a controlled environment for your seedlings. Sometimes plants that aren't as hardy do better when started indoors.

Which Seeds Should I Start Indoors?
Most root vegetables are better off being direct sown outdoors. The only root vegetable I actually start indoors is onions. Because I don't want to buy onion sets from the nursery, so I grow my own.
Cool-weather crops like lettuce, spinach, and other brassicas don't need to be started indoors as they can handle the cooler temperatures of the soil.
However warm-weather crops such as peppers, tomatoes, cucumbers, even melons, and some squash do really well when started indoors.
The seeds you start indoors will depend upon your zone and window of warm weather. If you have a really long growing period, then you may be able to direct sow all of your plants. If you live in the Pacific Northwest, like me, you'll need to start some of your plants indoors in order to even get a harvest.

When to Start Seeds Indoors

Seeds always germinate better when they're outside getting direct sunlight. But if the temperatures aren't right yet, this is why we start them inside.

You don't want to start your seeds too soon or too late (it's kind of like Goldilocks, finding the perfect time for your garden). This start date is always dependent on when your last frost is in the Spring and when your first frost is in the fall.

Once you know your first frost date (both my books, The Family Garden Plan or The Family Garden Planner have charts to help with this planning), you can count backward based on the number of weeks you should start those plants indoors to know when to get them going.

For example:

• Onions – If you want to start onions from seeds and not purchase onion sets, you'll need to start them first, 10 weeks before your last average frost date.

• Peppers and tomatoes are about 4 to 8 weeks before your last average frost, if you live where it's cooler, I recommend 8 weeks.

Micro-Climates

It's important to note that even with knowing your gardening zone and the general rule of thumb on your first and last average frost dates, there are always micro-climates in regions. I recommend asking around to your neighbors to get the best idea.

For example: According to the data I'm in gardening zone 7B and my last average frost date is April 29th and the first average frost date is October 14th (click here to find yours by your zip code).

But, I live up in the foothills and we have more extreme temperatures, which means we'll have a sneak frost sometimes come in as early as September 20th and we never plant our warm weather plants out in the spring until May 20th or later (depending upon the weather that year).

We usually don't plant until 1-2 weeks AFTER your last frost date because, typically speaking, your soil temperature is about 10 degrees cooler than the temperature outside. Talk to local gardeners in your area to find out when they plant, but the average frost dates and gardening zone give you a good rule of thumb.

What is Needed to Start Seeds Indoors

To successfully germinate and grow seeds indoors you'll need the proper combination of a few things:

• Seeds (Where to Buy Heirloom Seeds & My Favorite Seed Companies)
• Soil
• Containers
• Heat
• Water
• Light

Choosing The Right Soil

Young plants are just like infants, they're more susceptible to disease and getting sick. If you use dirt straight from your garden you're introducing any disease, bacteria, or fungus that's in that soil to a baby without an immune system.

This is often the cause of dampening off (a form of blight on seedlings) one of the biggest culprits of why seedlings die.

You're also bringing insects and their eggs into the house or gardening shed. Which you don't want on your baby seedlings or flying and crawling about your living room (which is where I seed start).

How to choose the right soil:

1. Purchase potting soil. I only use organic potting soil, it's already at the perfect mix and has been sterilized to kill any disease and/or fungus.

2. Bake your dirt at 200 degrees until it reaches 180 internally. This is your at home version but I'll be honest, I don't want to deal with trays and trays of dirt in my oven.

3. Make a mix of equal parts compost, top soil, and sand. Again, you'll need to sterilize it, but this is a DIY homemade way of making your own.

Seed Starting Containers

Too often people want to germinate their seeds in a larger pot so they don't have to transplant their seedlings into a larger pot before moving outdoors. However, seeds germinate best under a controlled environment and it's much easier to control the soil temperature of a small container of soil than if it's a big deep container.

I like to use plastic clamshell egg cartons because the holes are nice and small, plus when I close the lid the container acts like a natural greenhouse keeping the moisture and heat in.

Whether you repurpose an egg carton or buy the seed starting pods, you'll want to be sure there's a way to cover the container until germination is complete. For more seed starting containers, ideas and suggestions read What Are the Best Seed Starting Containers

Heat & Soil Temperature

The big question I get asked all the time is whether or not people need to buy seed starting mats in order to successfully start plants indoors. In my experience, the answer is no.

Even with my house temperatures dipping down in the mid to low 60's at night, I've never actually needed to use a seedling mat to successfully germinate seeds. However, if your house (or where you're starting your seeds) tends to be on the cooler side, a seed starting mat may be necessary.

Each seed has a unique temperature for germination, so check your seed packets to know how warm your soil temps need to be.

For warm-weather crops, your soil must be at least 60 degrees Fahrenheit or warmer for the seeds to germinate. The general rule of thumb is between 60 to 70 degrees, though your hot peppers may need to be 75+ for best germination rates.

You can use a soil thermometer or go by the average temperature in your home. I don't use seedling mats or hot pads and the temperature in our living room is usually 65 to 75 degrees (cooler in the early morning before the fire is built back up) and I haven't had an ounce of a problem getting my seeds to germinate.

Once seeds have sprouted, you want to make sure the overnight lows of the plants get below 50 degrees. So if your seedlings are sitting by a windowsill, you'll want to move them away from the window overnight so they don't get too cold by the window.

Adequate Water

This is where people often get into trouble. When both germinating seeds and watering your seedlings once they've sprouted you don't want the soil to be soaking wet, but you also don't want the soil to get too dry.

Watering for Germination

It's important to keep the soil consistently damp for the first 10+ days until the seeds have germinated. To keep the soil damp it's best to use a container that can create a greenhouse-like environment. This is why I like to use the plastic egg clamshell containers!

Depending on the seed, they typically take between 7-14 days to germinate. The closer you can get to the proper soil temperature and consistent moisture, the faster they'll germinate.

Watering Seedlings

You don't want the soil to continue to stay damp like you do when seeds are germinating, otherwise, you'll get rot and you may see some mold.

It's a good idea to allow the top surface of the soil to dry out between waterings because this forces the roots to go down in search of water.

I like to test the soil simply by touching it with my finger or pulling back the top layer of soil to make sure the soil is damp underneath.

Adequate Light

Plants don't actually need sunlight when they're germinating, so you don't need to have a sunny windowsill, or grow lights on during the first week or two. However, once they've germinated you have two options for light:

1. Sunny windowsill
2. Grow light

For most people, especially if you live in a northern climate, a sunny windowsill will not provide enough hours of daylight for your plants. You'll need 6 to 8 hours of direct sunlight for plants a day.
When using artificial light you also want to be sure the lightbulb is full-spectrum.
Because it's not as strong as the actual sun, you're going to need to leave the lights on for 16 to 18 hours every day. I turn mine on when I get up, first thing in the morning, and off when I go to bed.
I've had this same grow light for six years, including the same bulbs click here for my favorite grow light.

Proximity of Plants to Light
When using artificial light to grow your seedlings, it needs to be within 2-3 inches above the plant. Leggy and tall plants occur when plants aren't getting enough light, so they're stretching to reach the light source.
Most people don't give their plants enough light because they read the seed packets which tell you the plants need 6-8 hours a day, but that's full sunlight during the middle of summer, not an artificial light bulb.

How to Start Seeds Indoors

1. Once you have all the materials needed, you'll want to fill your containers about 2/3 full with potting soil.
2. Place 2-3 seeds per pot on top of the soil. A general rule of thumb is to plant twice as deep as the seed is wide. I find for smaller seeds when I water, this pushes the seeds under the soil just enough.
3. Add a small layer of soil over the seeds, if needed to be sure all seeds are covered.
4. Mist soil with a spray bottle filled with warm water.
5. Cover the container.
6. Water 1-2 times daily for 7-10 days, keeping the soil consistently moist until seeds germinate.

Germination Test

If you happen to have some really old seed packets and you're not sure if the seeds are viable (meaning you're not sure if they'll actually grow), you can do a paper towel germination test.
This test isn't actually for germinating plants, but just to test the viability of your seeds.
To test your seeds:
1. Take a damp paper towel and place 5-10 seeds on top.
2. Cover the seeds with another damp paper towel.
3. Keep the paper towel damp for 5-7 days, checking every few days to see how many of the seeds germinate.
This will give you a good indication of how many seeds you should sow in order to get the number of plants you want.
For example, if you want to grow 10 tomato plants in your garden, it's not a good idea to only plant 10 tomato seeds. Seeds are very inexpensive, so it's always better to plant more seeds (double or triple) than you'll need. That way you can keep the best, most robust-looking seedlings and give away what you don't need.

Caring for Plant Starts Indoors

Continue to give your plants adequate light and water for the duration of their life. Even once transplanted outdoors, you'll want to be sure they're planted in a location where they'll get enough sunlight.
After four weeks of growing, your plants will need additional care such as feedings, toughening up, and eventually hardening off before transplanting outdoors.

Feeding Your Plants

Once your seeds have germinated you will see the first set of leaves, these aren't actually their "true" leaves. These tiny leaves actually contain nutrition from the seed that feeds the seedling until the roots are big and strong enough to get nutrients from the soil (about the first 4 weeks).
The second set of leaves that appear are the plant's "true" leaves and they will actually resemble the leaves of the mature plant.

Once these "true" leaves appear it's critical to get the plants under a grow light and make sure the plant has enough nutrients (if you used an all-purpose potting soil, it's likely your plants will have what they need for the first 4 weeks).
At 4 weeks you'll want to start feeding your plants a bit of fertilizer. To do this I mix together some fish fertilizer with my water and water with it every week or even every other week to make sure the plants have enough nutrition.

Mimic Nature

Many people start seeds indoors but when they go to plant them outdoors they die. There are two reasons for this, you haven't hardened off your plants and you haven't toughened up your plants.
We'll cover hardening off next, but in nature, seedlings have rain hitting them, wind blowing around them, and they're constantly moving following the sun and being pushed around from airflow.
All of this helps the stem and the roots go deep. It develops a strong plant. Like working out, your muscles don't grow unless they're forced to work and move against something.

Mimic nature for your seedlings:
1. You can place a fan on the plants periodically to mimic the wind.
2. Run your hand over the leaves and the tops of the plant. This serves the same purpose as the fan but without the electricity and the fan. Whenever you walk by or turn on or off the light, run your hands over the plants.
3. Use a spray bottle periodically to mimic rain. This also helps if you have low humidity in your area and means you don't have to water as often. I haven't experienced any negative effects, even doing this on my tomato plants.
Make sure your plants dry fully between misting sessions, as well as the top part of the soil, after the seedlings have sprouted (germinated) and have their first sets of leaves.

Selecting & Saving Seed

Healthy and good quality seeds are the roots of a healthy crop. The seeds that are used to cultivate new crops have to be selected very carefully and of high quality. The good quality seeds can either be bought from different sources or farmers can produce by their own. The selection of seeds is used to improve the quality of yields. There are several diseases that are transmitted via the seeds. If the selected seeds are from the infected fields then the seed-borne diseases will cause severe problems in the agricultural process. Thus, always obtain seeds from healthy plants. Along with the diseases free and healthy seeds, farmers also need to check the germination period of the seeds, nutrients required and other benefits in terms of yield and finance. Overall, selecting good quality seeds are essential for growing strong and healthy crops.

When compared to the history of agricultural being able to order seeds each year is a relatively new thing. For thousands of years, seed saving was simply a part of gardening. This family tradition created the heirlooms we know today and adapted varieties to specific local conditions.

Whenever we save seeds we're making decisions about what future generations of plants will look like. Saving seed for next year takes careful consideration. Especially with easy to save species like potatoes or garlic it can be tempting to plant whatever is leftover or whatever you don't want to cook with but it's important to remember this will affect future harvests. Take a look at some of the traits you should consider when saving seed.

Vigor

This is a plant's ability to germinate well and quickly grow into a healthy, productive plant.

Earliness

Even if you have a relatively long season, selecting for early production may be desirable. Plants that mature quickly can allow you to enjoy produce earlier in the year, avoid the intense heat of late summer, avoid certain pests, or allow you to harvest multiple crops.

Trueness-to-type

If you're trying to help preserve an heirloom variety you want to be sure to save only save seeds from plants that are true to type, displaying the characteristics unique to that variety.

Disease and Insect Resistance

Saving seed from year to year can adapt a variety to better withstand your local pests and diseases.

Tolerance to Drought or Excess Moisture

This trait is another way you can adapt a variety to your specific garden location.

Stockiness

Tall, spindly plants can be prone to lodging and other problems and often need additional trellising. Stocky plants are often healthy plants. Note, this trait can also be affected by nutrient availability and how closely you space your plants.

Hardiness

Particularly for crops grown in early spring or late fall you want to select for plants that withstand cold temperatures.

Lateness to Bolt

Avoid saving seed from the first plants that go to seed. Selecting for those that bolt later will increase your harvest period.
Color
It may not seem as important as disease resistance or flavor but often times their unique color is why people fall in love with an heirloom variety. Selecting for the most intense Purple Dragon Carrots is part of what makes the variety special.

Uniformity or Lack of It

This trait will depend upon your variety. You may want all of your green arrow pea plants to be a uniform height but obviously, you don't want your rainbow swiss chard to have a uniform color.

Flavor

One of the best things about heirlooms is their amazing flavor! Always take this into consideration in your seed saving endeavors. Life is too short for tasteless vegetables.

Flesh Characteristics

This trait will largely depend on a variety's purpose. Tomatoes like Principe Borghese which were bred for drying should have much less moisture than slicers like Radiator Charlie's Mortgage Lifter.

Size and Shape

If you love stuffed jalapenos save seed from the biggest peppers. This is another great way to make a variety work for you.

Productivity

Don't eat your biggest, best cabbages even though it's tempting. Let those go to seed so that in a few years more of your cabbages will resemble the best.

Storage Ability

Although it may seem less important in modern times when everyone has a fridge and freezer, storage ability should still be considered particularly in plants like storage tomatoes, pumpkins, winter squash, sweet potatoes, etc.

There are so many qualities to consider when selecting seeds it can be tough to keep track. One thing you can do to mark a specific plant is to loosely tie a bright colored piece of yarn around it. This will allow you to remember which plants displayed traits like vigor when it's time to harvest seed. Be sure to keep an eye on the yarn as the plant grows so it doesn't get too tight and harm the plant. Alternatively, you can place small stakes in front of plants to mark them.

Growing a Plant to Save Its Seed Is Different Than Growing It to Eat

And usually, you're not getting both. In order for a plant like lettuce to produce seed, you must wait for it to send up its gangly flower stalks, which eventually produce tiny seed pods. By this time. the lettuce leaves are becoming yellow, shriveled, and bitter. It's the same with most crops – you don't get to eat it and save the seed; it's either one or the other. The good news is that a single plant produces many seeds. So you usually need to grow only a few extra for seed-saving purposes.

Don't Bother Saving Seeds of Hybrid Varieties

Seeds denoted on the package as "F1" are hybrids, meaning two varieties have been bred with one another (cross-pollinated, that is) to produce a third variety with a combination of traits from each "parent." If you were to save seed from this hybrid offspring and plant it, each seed would grow into a plant with a random combination of the traits found in the gene pool of the original parents, which rarely produces something you'd want to eat. The only way to reproduce the hybrid "true-to-type," as plant breeders say, is to cross the two original parents. That's a big part of why most seed savers stick with old-fashioned heirloom varieties, which by definition are not hybrids.

Save Seeds from the Best Plants

To save seed is to participate in the process of natural selection. If you save seed only from the biggest tomato of the bunch and replant them year after year, you'll eventually end up with seeds that produce plants on which all the tomatoes are bigger. The same holds true for almost any other trait. Want tomatoes that ripen earlier? Save seed from the first fruits to ripen each year. Want disease resistant plants? Then definitely don't save seed from those that are disease-infested. This is essentially what professional plant breeders do. You don't need to get too scientific about it, but as a rule of thumb, only save seed from your healthiest, most robust, tastiest plants.

Seed Saving Can Be Tedious

Bean seeds are big and easy to remove from their pods. But this is the exception rather than the rule. Carrot seeds, for example, are no bigger than a baby flea and easily disappear into the nearest crack or cranny as you try to knock them loose from their seed heads. Plants hold their seeds in an array of husks, pods, capsules, and other coverings, which are often not easily removed. This process varies depending on the species in question, but typically involves threshing (separating the seed from the plant) and winnowing (separating the seed from its hull). If you're collecting only a very small quantity of seed, you'll probably perform these tedious tasks by hand, but specialized tools are available for processing larger quantities.

Seed Saving Can Be Stinky

Seed that develops in a wet, fleshy fruit (tomatoes, melons, and cucumbers, for example), as opposed to a dry seedhead or pod (the case with most greens, herbs and legumes), often requires extra steps to extract. Such seed is typically encased in a gooey substance, from which it is not easily removed. The best way to remove the goo, as it turns out, is to put it in a jar or bucket with a bit of water and let the concoction rot for a bit. The fermentation process dissolves the goo and improves the germination rate of the seed. You then strain the seeds from the stinky liquid and dry them.

Seeds of Some Crops Are Easier to Save Than Others

Seeds are the products of pollination, the botanical version of sex. Some crops are self-pollinators, which means individual plants are fertilized by their own pollen. These crops, including beans, peas, tomatoes, peppers, and cauliflower, are among the easiest to save because you don't need special botanical knowledge to ensure that the seeds grow out true-to-type.

Plant Sex Makes Things Complicated

It's with cross-pollinating crops – those that need pollen from a neighboring plant in order to set seed – where things get complicated. This group includes cucumbers, corn, squash, pumpkins, and melons. If you have more than one variety of the same cross-pollinating vegetable (a butternut squash and acorn squash, let's say) growing in close proximity, pollen from one will inevitably end up in the flowers of the other, resulting in seeds that are a mutant hybrid of both varieties. Seed savers employ various strategies to prevent this, ranging from growing different varieties on opposite ends of their property (pollen only travels so far on the wind or via insects) to placing plastic bags over some flowers to exclude unwanted pollen (you must then use a paintbrush to pollinate them with pollen from the same variety). Another option? Simply grow only one variety at a time of these particular crops.

Seeds Aren't Viable Until Fully Ripe
Just like picking the perfect tomato, you have to wait until seed is fully ripe before you harvest it – if picked from the plant too soon, the seed will not germinate. As explained above, optimal seed maturity is usually later than optimal crop maturity. Bean and pea seeds are not ready until the pod is brown, dry, and beginning to split open. This is true of any seed that grows in a pod, which includes most greens. Corn seed should be allowed to dry on the cob in the field. Some vegetables, including cucumbers and eggplant, should not be picked for seed until they are overripe and beginning to shrivel up and rot.

Well-Dried Seed Is Viable Seed
In order to preserve seed for future plantings, it must be thoroughly dry. Drying out is essentially the final stage of ripening, and ensures that the seed does not become moldy while you're waiting to plant it next year. Wet seed, once it has been extracted from its fermented goo, must be spread out to dry on screens in a warm location, ideally with a light breeze from a fan to hasten the process. Most other types of seed my be dried while still on the plant, but if the weather turns wet and cool before that can occur, you'll need to bring them indoors to finish the process. To determine if seed is sufficiently dry, push a fingernail into it – if it gives, it's not yet ready.

Proper Storage Is Important
Dried seed should be placed in paper envelopes or seed packets labeled with the name of the variety and the date it was harvested. To ensure longevity, keep the seed packets in mason jars in a cool dark place. Any seed stored this way should remain viable for at least a few years, though some crops may keep for a decade or more.

Season Extension

In agriculture, season extension refers to anything that allows a crop to be cultivated beyond its normal outdoor growing season. Advantages Possible year-round income Retention of old customers Gain in new customers Higher prices Higher yields Better quality Extended employment for workers Disadvantages No break in yearly work schedule Increased management demands Higher production costs Plastic disposal problems How does season extension contribute to sustainability? "…to make a real difference in creating a local food system, local growers need to be able to continue supplying "fresh" food through the winter months…[and] to do that without markedly increasing our expenses or our consumption of non-renewable resources". – Eliot Coleman, The New Organic Grower Thermodynamics and Properties of Plants We can extend the growing and harvest season for crops through techniques that evolve from two primary strategic goals: Protecting crops from damage from extremes of heat or cold Enhancing the growth of crops for quicker maturity and higher quality under adverse weather conditions Often one technique will affect more than one strategy. For example, a raised bed will dry faster and warm up sooner in spring, but will therefore require more attention to irrigation needs, and may develop higher-than-desirable soil temperatures sooner than flat ground when summer arrives. It can also cool faster than flat ground. Another example: A row cover may protect a crop from a frost, but can also prevent the crop from developing as much hardiness as an uncovered crop, due to the artificial mild climate under the cover. Growers are much more likely to succeed in their efforts to extend the growing season for crops if they understand some basic principles about heat and cold, about how plants respond to thermal changes, and how various landscape features and protective materials influence the thermal environment of plants. Controlling the flow of heat The ground is a huge reservoir of heat. It's the heat radiating from the soil that protects crops at night when you cover them, or protects the lower part of plants that have a good canopy of leaves. As the ground gets colder in late fall, it radiates less heat. Thus the leaf canopy or added row covers will give less protection as the ground cools. The reverse is true as the soil temperature increases in spring. Wet ground conducts and radiates more heat than dry ground. Lighter, sandier soils and any soil in a raised bed will dry out, warm up, and become workable earlier in the spring. Adding organic matter to clay soils can improve drainage; this will also darken any soil so it will absorb more heat. Dark mulches, such as black plastic, also raise soil temps. Mulches will insulate the ground and allow less heat out at night; thus crops will be colder on mulched ground. Clouds form a "blanket" that slows radiant heat loss from the earth. Temperatures can drop as much as 5-10° F within an hour after the sky clears at night. The arrival of cloud cover will often raise temperatures and save crops on a frosty night. Water can store a lot of heat and can give it up relatively quickly. This enables overhead irrigation to protect a crop from frost. It also creates warmer microclimates near ponds or other bodies of water. Water also absorbs heat quickly, and evaporating water removes heat from its environment. Thus, overhead irrigation on a hot day can cool heat-sensitive crops or aid germination of heat sensitive seeds. Land sloping to the south will stay warmer in the late fall and warm up sooner in the early spring. Land sloping east will warm sooner in the morning; to the west it stays warmer in the evening. Cold air is heavier than warm air, and will slide down slopes to settle in flat areas or hollows, often referred to as "frost pockets". On a calm night, the warmest "microclimate" in a given area will often be near the top of a slope. Objects in a landscape, such as buildings or created windbreaks, can influence the movement of cold air. Windbreaks uphill from a crop can protect from frost; downhill from a crop may cause trapping of cold air. A forest surrounding a smaller, level field on several sides may also keep cold air from reaching the crops. Windbreaks, such as buildings, hedges, fencerows, or woods, can influence microclimates in other ways. They can create a useful "microclimate", where solar gain can accumulate in daytime; wind can stress plants by accelerating evaporation from leaves; it can also push cold air deeper into the canopy of the crop or through a row cover. The "wind chill factor" means: moving air extracts heat from objects faster than still air. Remember: Weather forecasts are usually for urban areas; temperatures on clear nights can be as much as 10 degrees lower in rural frost pockets than what is stated in the media for a general local area. Hardiness: Which Crops Have How Much, and Under What Conditions? If the weather gets colder gradually, plants will develop hardiness, and be less damaged by extreme cold. If extreme cold immediately follows a warm spell, it will do much more damage to most crops. Moderate wind also helps to harden crops, both in the field and in seed flats. "Soft" transplants, freshly transplanted to the field, are more vulnerable to frost damage than hardened plants. Higher nitrogen levels will keep plants soft, less hardy. Various stresses and health conditions can also lessen hardiness. Each crop, when thoroughly hardened, has a typical threshold temperature, below which it will usually be damaged. Within each crop, varieties will vary in their hardiness, hence the range of temperatures given in the following table. Consult seed catalog variety descriptions for mention of degree of hardiness. The table below is based on Doug Jones' personal experience. Threshold Damage Temperatures for Common Crops TENDER CROPS (crops originally from the tropics): All are damaged at 31-32° F. These include all the nightshades – tomatoes, peppers, etc. (young potatoes can recover from moderate frost), cucurbits (squashes, cukes, melons, etc), beans, basil, sweet potatoes, and corn (can also recover). COOL SEASON CROPS: Most crops below will survive at temperatures lower than the indicated range. If winter survival is the main goal, some damage is acceptable; the plants will usually recover as the weather warms up.

VERY HARDY Leeks (5-20° F) Parsnips (28° F, tops 15° F, roots 0° F) Spinach (8-12°) HARDY Cabbage (12-20° F) Broccoli (18-22° F) Brussels Sprouts (15-20° F) Rutabaga (10-15° F) Kale (10-15° F) Mustard (10-15° F) SEMI-HARDY Lettuce (20-25° F) Cauliflower (25-28° F) Carrot (tops 10° F, roots 30° F) Beets, Chard (15-25° F) Peas (18-25° F, blossoms 30° F)

Slow-Growing Cool Season Crops – a Special Challenge in the Piedmont In the Piedmont, we have two seasons of moderate weather enjoyed by cool weather crops (60-75° days): spring and fall. These seasons are often less than 60 days long. The quick-maturing crops are relatively easy to fit into this window: lettuce, spinach, mustard greens, radishes, spring peas, fall carrots. Using transplants helps to fit the cole crops into spring and fall, though the fall seedlings need to be started in a shaded location, and the spring crop in a greenhouse or cold frame. Very slow crops, like parsnips, Brussels sprouts, and celery are more difficult. Row covers can help to get these crops started extra early in spring, so they can finish before hot weather degrades them. Covers can help the fall crop mature without damage. Temperature Extremes Affecting Germination and Seedling Vigor One limitation of early spring planting is the negative effect of cold soil on germination. Seedlings that emerge slowly are much more vulnerable to attack by soil pathogens. Clear plastic laid directly on a seedbed can help seeds germinate quickly, provided the soil doesn't overheat. Conversely, hot soil inhibits germination of some fall crops, particularly spinach, lettuce, parsnips, beets. Use mulches, shading, and/or micro-sprinklers to cool the soil and crop. Days required for vegetable seedling emergence at various soil temperatures for seeds planted ½ inch deep

Soil Temperature (° F)

Crop	41	50	60	68	77	86	95
Cool Season Broccoli	28	20	10	7	5	4	–
Cabbage	24	15	9	6	5	4	–
Cauliflower	30	20	10	6	5	5	–
Onion	31	13	7	5	4	4	13
Pea	36	14	9	8	6	6	–
Spinach	23	12	7	6	5	6	NG
Warm Season Snap Beans	NG	NG	16	11	8	6	6
Eggplant	–	–	27	13	8	5	7
Muskmelon	–	–	NG	8	4	3	–
Pepper	NG	NG	25	13	8	8	9
Sweet Corn	NG	22	12	7	4	4	3
Tomato	NG	43	14	8	6	6	9

There are lots of things you can do to modify the crop microclimate that do not involve covers or structures. Some of these strategies require long-term planning. Site Selection Land with a south-facing slope will stay warmer in the late fall and warm up sooner in the early spring. A site at the top of the slope, with unimpeded air drainage down the hill, would be ideal for maximizing season extension. Soil and Moisture Management Adding organic matter, tillage, raised beds, improve drainage Cultivar Selection Cultivar selection is an important strategy for achieving season extension. The number of days from planting to maturity varies from cultivar to cultivar, and some cultivars germinate better in cool soil than others. Staggered planting dates can be combined with the use of cultivars spanning a range of maturity dates to greatly extend the harvest season for any one crop. Heat-tolerant varieties can help extend the season of certain crops. See ATTRA publication Scheduling Vegetable Plantings for Continuous Harvest Transplants Use of transplants is a key season extension technique. They can provide a 3-4 week head start on the season. Irrigation Increasing or decreasing soil water content can enable tillage operations, prevent waterlogging of the root zone and/or aid germination. Overhead irrigation can be used to protect crops from frost. Micro-sprinklers can be used to cool them. Windbreaks Use of windbreaks can result in increased yield and earlier crop production by providing wind protection. Young plants are most susceptible to wind damage and "sand blasting". Rye strips between rows can provide protection from wind and wind-borne sand. Windbreaks can improve early plant growth and earlier crop production, particularly with melons, cucumbers, squash, peppers, eggplant, tomatoes, and okra. The major benefit of a windbreak is improved use of moisture. Reducing the wind speed reaching the crop reduces both the direct evaporation from the soil and the moisture transpired from the crop. This moisture advantage also improves conditions for seed germination. Seeds germinate more rapidly and young plants put down roots more quickly. Improved moisture conditions continue to enhance crop growth and development throughout the growing season. The type and height of the windbreak determine its effectiveness. Windbreaks can be living or non-living. Rye strips are suggested for intensive vegetable production based on economics. In general, the windbreak has a significant effect on the crop at a distance of 8 times the height of the windbreak. A rye strip 3 feet tall will protect vegetables up to 24 feet away. The windbreak should be planted perpendicular to the prevailing wind direction. Rye strips should be planted from September through October to obtain good plant establishment and to provide adequate time for plant growth prior to beginning the next production season. Fabrics and Structures for Season Extension Plastic Mulch Plastic mulches have been used commercially for the production of vegetables since the early 1960s. Plastic mulches provide many positive advantages for the user, such as increased yields, earlier maturing crops, crops of higher quality, enhanced insect management, and weed control. They also allow other components, such as drip irrigation, to achieve maximum efficiency. Although a variety of vegetables can be grown successfully using plastic mulches, muskmelons, tomatoes, peppers, cucumbers, squash, eggplant, watermelons, and okra have shown the most significant responses. Production of strawberries and cut flowers is also greatly improved by the use of plastic mulch. Plastic mulches directly impact the microclimate around the plant by modifying the radiation budget (absorbitivity vs. reflectivity) of the surface and decreasing the soil water loss. The color of a mulch largely determines its energy-radiating behavior and its influence on the microclimate around a vegetable plant. Color affects the surface temperature of the mulch and the underlying soil temperature. Another important factor is the degree of contact between the mulch and soil or by not being taut, often quantified as a thermal contact resistance, will greatly influence the performance of a mulch. If an air space is created between the plastic mulch and the soil by a rough soil surface, soil warming can be less effective than would be expected from a particular mulch. Black plastic mulch – the predominant color used in vegetable production is an opaque blackbody absorber and radiator. Much of the solar energy absorbed by black plastic mulch is lost to the atmosphere through radiation and forced convection. The efficiency with which black mulch increases soil temperature can be improved by optimizing conditions for transferring heat from the mulch to the soil. Because thermal conductivity of the soil is high relative to that of air, much of the energy absorbed by black plastic can be transferred to the soil by conduction if contact is good between the plastic mulch and the soil surface. Soil temperatures under black plastic mulch during the daytime are generally 5° F higher at a 2-inch depth and 3° F higher at a 4-inch depth compared to those of bare soil. Clear plastic mulch – absorbs little solar radiation but transmits 85% to 95%, with relative transmission depending on the thickness and degree of opacity of the polyethylene. The undersurface of clear plastic mulch usually is covered with condensed water droplets. This water is transparent to incoming shortwave radiation but is opaque to outgoing longwave infrared radiation, so much of the heat lost to the atmosphere from a bare soil by infrared radiation is retained by clear plastic mulch. Thus, daytime soil temperatures under clear plastic mulch are generally 8 to 14° F higher at a 2-inch depth and 6 to 9° F higher at a 4-inch depth compared to those of bare soil. Clear plastic mulches generally are used in the cooler regions of the United States, such as the New England states. Weeds can be a big problem under clear mulch. White, white-on-black, or silver reflecting mulch – can result in a slight decrease in soil temperature (-2° F at 1-inch depth or -0.7° F at a 4-inch depth) compared to bare soil, because they reflect back into the plant canopy most of the incoming solar radiation. These mulches can be used to establish a crop when soil temperatures are high and any reduction in soil temperatures is beneficial. Infrared-transmitting mulch – These mulches provide the weed control properties of black mulch but are intermediate between black and clear mulch in terms of increasing soil temperature. The color of these mulches can be blue-green or brown. These mulches warm up the soil like clear mulch but without the accompanying weed problem. Red mulch – performs like black mulch, warming the soil, controlling weeds, and conserving moisture, with one important difference. In Pennsylvania experiments, tomato crops responded with an average 12% increase in marketable fruit yield over a 3-

year period. There appears to be a reduction in the incidence of early blight in plants grown on red mulch, compared with plants grown on black mulch. When environmental conditions for plant growth are ideal, tomato response to red mulch is minimal. Other crops that may respond with higher yields include eggplant, peppers, melons, and strawberries. Source: Center for Plasticulture at Penn State Biodegradable mulch Biodegradable plastics are made with starches from plants such as corn, wheat, and potatoes. They are broken down by microbes. Biodegradable plastics currently on the market are more expensive than traditional plastics, but the lower price of traditional plastics does not reflect their true environmental cost. Field trials in Australia using biodegradable mulch on tomato and pepper crops have shown it performs just as well as polyethylene film, and it can simply be plowed into the ground after harvest. Researchers with Cornell University also found that biodegradable mulches supported good yields, but the films they used are not yet commercially available in the U.S. Bio-Film is the first gardening film for the U.S. market. Made from cornstarch and other renewable resources, it is 100% biodegradable. Bio-Film is available from Dirt Works in Vermont. Paper mulch can provide benefits similar to plastic and is also biodegradable. An innovative group in Virginia carried out on-farm experiments to explore alternatives to plastic. They compared soil temperatures and tomato growth using various mulches, including black plastic, Planters Paper, and recycled kraft paper. Plastic, paper, and organic mulches all improved total yields of tomatoes grown in the trials, when compared with tomatoes grown on bare soil. Recycled kraft paper is available in large rolls at low cost. Participants in the experiment were concerned that it would break down too quickly. To retard degradation, they oiled the paper. This resulted in rather transparent mulch and, as with clear plastic, soil temperatures were higher than under black plastic. Weeds also grew well under the transparent mulch. To reduce weed growth and to keep the soil from becoming too hot, the experimenters put hay over the oiled paper several weeks after it was laid. Planters Paper is a commercially available paper mulch designed as an alternative to plastic. It comprises most of the benefits of black plastic film and has other advantages. It is porous to water. Left in the soil, or tilled in after the growing season, it will degrade. Tomatoes grown with this mulch showed yields and earliness similar to tomatoes grown in black polyethylene mulch, even though the latter resulted in slightly higher soil temperatures. Planters Paper, however, is considerably more expensive than black plastic. It does not have the stretchability of plastic, and it tends to degrade prematurely along the edges where it is secured with a layer of soil. The paper is then subject to being lifted by the wind. Rebar, old pipe, or stones — rather than soil — can be used to secure the edges.

Row Covers Two main types of row covers: floating row covers – lie directly over the crop and may cover multiple rows Hoop-supported row covers – sometimes referred to as low tunnels, they generally cover a single row There are also two basic types of row cover materials: clear polyethylene spunbonded polyester or polypropylene All of these are available in varying thicknesses, weights, widths, and lengths. Crop distinctions such as temperature sensitivity, pollination methods, and growth habit dictate the type of row cover that is best to use. Spunbonded covers are comprised of a thin mesh of white synthetic fibers which entrap heat and serve as a barrier to wind, insects, and varmints. Water from rain or overhead irrigation freely passes through. The weight of these covers ranges from 0.3 to about 2.0 oz/sq yd. The lightest covers are used primarily for insect exclusion while the heaviest of the covers are used for frost protection. The most common medium weights are 0.5 to 1.25 oz/sq yd and are best for season extension for such crops as melons, cucumbers, squash, lettuces, peas, carrots, radishes, potatoes, sweet corn, strawberries, and cut flowers. With covers under 0.5 oz there is minimal heat retention at night; and over 1.75 oz, there is a significant reduction in light transmission. The heavy covers are used for nighttime frost protection only since they do not transmit sufficient light for optimum crop growth. Spunbonded row covers in the 0.5 to 1.25 oz range provide 2-4° F frost protection in the spring. In the fall, there is more protection because there is a larger reservoir of heat in the soil in the fall than in the spring. Floating row covers Floating covers require much less installation labor than hoop supported covers. The wider and longer the covers, the less labor required per unit area since only the edges are secured. These covers vary in width from 3 to 60 feet and up to over 2,000 feet long. Even though most crops can be grown without damage under floating covers, tomatoes and pepper are an exception. If the spunbonded material is not supported with wire hoops for these crops, flapping of the cover in the wind will damage the growing points of young plants. Also, with summer squash under windy conditions, many of the leaves might be broken by the cover. For these three crops, a series of strategically placed wire hoops will prevent crop damage. Weed control can be a significant problem under row covers unless they are used in combination with plastic or some other type of mulch. Hoop-supported row covers Hoop-supported row covers are often called low tunnels. They offer many of the same benefits as floating row covers, but are not permeable to air or water and are more labor-intensive. There are several types of low tunnels. Slitted row covers have pre-cut slits that provide a way for excessive hot air to escape. At night the slits remain closed, reducing the rate of convective heat loss and helping to maintain higher temperatures inside the tunnel. Punched row covers have small holes punched about 4 inches apart to ventilate hot air. The punched covers trap more heat than the slitted tunnels. They are best for northern areas and must be managed carefully to avoid overheating crops on bright days. They are useful for peppers, tomatoes, eggplant, most cucurbits, and other warm-season crops that grow upright. Additional Observations by Doug Jones: In the protected, wind-free, slightly warmer environment under the cover, crops will grow faster during cold weather. Ironically, this effect also prevents plants from developing as much hardiness as equivalent crops with no protection! To deal with this, I add a second cover when extreme cold is forecast. In the Piedmont, many cool-season crops can be brought through the winter under an adequate thickness of fabric (medium weight fabric: add a second layer). Crops actually grow a little during winter warm spells. High (hoophouse) or low tunnels with a single layer of clear plastic get much warmer on sunny winter days than fabric covered beds, so crops will grow faster. But it gets about as cold at night in the hoop house as under the fabric covers. In the coldest weather I like to add a row cover to beds of semi-hardy crops inside the hoophouse. Clear plastic must be ventilated on calm, sunny days when ambient temperatures are above 50-60° F. Otherwise, the crop will be overheated. Plant leaves that touch the fabric usually get more damaged from extreme cold, and from strong winds whipping the cover back and forth. A tunnel structure of heavy (9 gauge) wire hoops will help to keep the fabric from touching the leaves of the crop. Concrete block wire ("ladder wire") makes a more stable tunnel structure. It's important to stretch the fabric very tight end-to-end, and use lots of pins, soil, or rocks to keep big winds from blowing your covers into the trees. Red plastic "Repins™" are very effective for pinning, as well as easy to pull up for access to the crop. Wire hoops are not sufficient to prevent heavy snow from collapsing the cover. Sometimes I just take the cover off when a big snow is forecast, and actually get less damage than under the collapsed tunnels. Powdery snow itself is a good insulator. High Tunnels A high tunnel, or hoophouse, is basically an arched or hoop-shaped frame covered with clear plastic and high enough to stand in or drive a tractor through. Traditional high tunnels are completely solar-heated, without electricity for automated ventilation or heating systems. High tunnels have been used for many years throughout Europe, Asia, and the Middle East and they are rapidly gaining in popularity here. Several universities are now conducting high tunnel research. Benefits of high tunnels: Crops grown in hoophouses have higher quality and are larger than those grown in the field. Crops grown in hoophouses can hit the market early when prices are high and help to capture loyal customers for the entire season. Hoophouses allow certain crops to be grown throughout the winter, providing a continuous supply to markets the entire year. . The most critical components of a hoop house for strength are the end walls. Crops that have been grown in high tunnels include specialty cut flowers, lettuce and other greens, carrots, tomatoes, peppers, squash, melons, raspberries, strawberries, blueberries, and cherries. Although high tunnels provide a measure of protection from low temperatures, they are not frost protection systems in the same sense that greenhouses are. On average, tunnels permit planting about three weeks earlier than outdoor planting of warm season crops. They also can extend the season for about a month in the fall. Hoophouses are not difficult to build. The most common design uses galvanized metal bows attached to metal posts driven into the ground 4 feet apart — a traditional quonset style structure. Strength is important. Heavy-gauge galvanized steel pipe is best for hoops. Setting the hoops four feet rather than any further apart is also recommended. Growers in snowy climates might choose a peaked-roof structure instead of the quonset style. The most economical covering is 6-mil greenhouse grade, UV-treated polyethylene, which should last three to five years. Do not use plastic that is not UV-treated — "It will fall apart after half a season". Some growers use a double layer of plastic. Roll-up sides used on many high tunnels provide a simple and effective way to manage ventilation and control temperature. The edge of the plastic is taped to a one-inch pipe that runs the length of the tunnel. A sliding "T" handle is attached to the end of the pipe so that the plastic can be rolled up as high as the hip board. Ventilation is controlled by rolling up the sides to dispel the heat. Depending on temperature and wind factors, the two sides may be rolled up to different heights. Temperature management using the roll-up sides is critical. On sunny mornings, the sides must be rolled up to prevent a rapid rise in temperature. Tomato blossoms, as mentioned earlier, will be damaged when temperatures go above 86° F for a few hours. Even on cloudy days, rolling up the sides provides ventilation to help reduce humidity that could lead to disease problems. The sides should be rolled down in the evening until night temperatures reach 65° F. A maximum/minimum thermometer is a great aid in keeping tabs on temperature. Twenty feet by 96 feet is a size commonly used by market growers. This size allows efficient heating and cooling, efficient growing space, and adequate natural ventilation. Alison and Paul Wiediger of Au Naturel Farm in Kentucky also use a commercially available 20-by 96-foot high tunnel, in addition to two 21-by 60-foot tunnels. They think there is value in building as large a structure as is practical. They also find that plants close to the walls do not grow as well as the plants further from them. The larger the frame, the larger the percentage of effective growing area. And most growers want more, rather than less, space at the end of one growing season. Most of the growing in this tunnel will be in spring, fall, and winter when outside temperatures are cooler/colder. We believe that both the earth and the air within the tunnel act as heat sinks when the sun shines. At night, they give up that heat, and keep the plants safe. The smaller the structure, the smaller that temperature "flywheel" is, and the cooler the inside temperatures will be. The Wiedigers use a double layer of 6-mil, 4-year poly to cover their tunnels. A small fan blows air between the two layers to create an insulating barrier against the cold. Construction and management details can be found in their manual, Walking to Spring: Using High Tunnels to Grow Produce 52 Weeks a Year (See Resources handout). Source: ATTRA Season Extension Techniques for Market Gardeners Eliot Coleman popularized the concept of the mobile high tunnel in the U.S. The tunnels sit on railroad wheels and roll on wooden rails so they can be moved from one site to another. For example, the tunnel might be used to start a lettuce crop, and once the spring warmed up enough for the lettuce to thrive outdoors, the tunnel is moved to the second site where tomatoes are planted. Coleman's book Four-Season Harvest provides details on the mobile tunnel model. Haygrove Multibay Tunnel Systems With Haygrove tunnels, innovative growers are literally covering their fields to protect high-value crops from early and late frost, heavy rain, wind, hail, and disease. The frames also provide support for shade cloth and bird netting. The British company Haygrove was started in 1988 with a little more than two acres of strawberries in hoophouses. By 2002, Haygrove had expanded to nearly 250 acres of soft fruits, including strawberries, blackberries, red currants, and cherries, grown under plastic in England and eastern Europe. They also came up with a new design for multibay tunnels and sold 3,000 acres of tunnels throughout Europe. Haygrove tunnels are now being distributed and used in the U.S. Haygrove sells tunnels from 18 to 28 feet wide per bay, with a three bay minimum. There are no walls between bays. The total length and width can be whatever the grower desires. Company representative Ralph Cramer in Lancaster County, Pennsylvania, says he has seen tunnels as short as 65 feet and as long as 1,100 feet. They have been used to cover from 1/3 acre to 100 acres (of blueberries in California). Unlike greenhouses, Haygrove tunnels don't need to be built on flat ground, but can be built on slopes. Cramer says advantages of the systems include lower cost and better ventilation. One acre of Haygrove tunnels costs about 55 cents per square foot, or about $24,000. Local Example of Haygrove Tunnel System: Peregrine Farm Locally, Alex and Betsy Hitt of Peregrine Farm purchased a half an acre (two ¼ acre blocks) of Haygrove tunnels and first used them for the 2004 growing season. They had actually gotten a quote before the 2003 season began and the price scared them so they walked away. But then the summer of 2003 yielded a horrible tomato crop due to all the rain and disease so they decided to go for it. They had seen the Haygroves being used for cut flower production in California and Pennsylvania. They wanted them mainly as "dry houses" to keep the tomatoes dry and also for certain flowers that don't like their blooms getting wet (lisianthus, poppies, campanula, delphinium, etc.). Alex and Betsy already had 6 mobile high tunnels that worked great for tomatoes but they wanted something that could be moved with the rotation, cover lots of ground, and be well-suited for cover crops. They contacted Ralph Cramer of Cramer's Posie Patch, the east coast distributor for Haygrove. They ordered two ¼ acre blocks because that was the size of their rotation plots. Each bay is 24 feet wide. The Haygrove tunnels are shipped over from England. It takes about 8 weeks. The shipping cost alone is $4,000, whether the cargo container is full or not. If growers did a cooperative purchase they could reduce shipping costs. The total cost, including shipping, was $0.85/square foot. The plastic is supposed to last for at least two years; Alex will get at least three years. Construction of Haygroves: You have to be prepared for the delivery. You have to unload everything within two hours or you get charged Alex received over 5 tons of stuff! The roll of plastic itself weighed one ton. Need to have appropriate equipment ready for off-loading. The Hitts constructed the tunnels in October and covered them with plastic the following April. Haygrove sends someone to advise you on construction. They won't do any of the actual work but they are there to tell you how to do it. The Hitt's Haygrove rep spent an entire day with them and showed them how to lay out the field. It took two people 80 hours total to construct ½ acre of Haygroves. It took 4 people an additional hour to cover each bay in plastic. The Hitts use one ¼ acre block for cut flowers and one for heirloom tomatoes. Advantages of Haygroves Disease protection – Alex has a mysterious strain of anthracnose disease that attacks his tomatoes. This disease has even left our plant pathologists stumped. Under the Haygroves, the disease comes 3-4 weeks later, so instead of 5 weeks of production, he gets 8-9 weeks, and a 35% increase in income. At this rate, it will take three years to pay off the tunnels. They are still trying to work out which crops are best suited for the Haygroves. For some crops, it may increase disease. Dahlias had powdery mildew under the Haygroves. Improved quality – Alex believes they would be great for producing a superior quality melon since they would control the amount of rain which often dilutes the sugars and causes splitting. They grew melons in the Haygroves the first year and they did great until the plastic had to be removed for an impending hurricane and then the melons all exploded! Provides season extension – doesn't just affect temperature but also wind and moisture. The multiple huge bays means there's a large area of air under there that cools off slower at night. The Haygroves were originally designed in England to keep strawberries dry, so they are designed more for controlling rainfall rather than temperature. However, the Hitts did have a surprise frost hot after planting their tomatoes – the ground outside was totally white and the tomatoes inside the tunnels were fine. Disadvantages of Haygroves They are very difficult to construct if you have rocky soil, as the Hitts do. The legs are sunk 30 inches, and Alex had to use a jackhammer! This causes problems not only with initial installation but also makes it difficult to move them. To get around this problem, the Hitts are going to install another set of permanent legs, which the Haygroves can be moved to. This will allow them to have a three year rotation with the Haygroves. The Haygroves present a huge sail out in the field so you have to be prepared to vent them by pushing the plastic up very high. They are designed to withstand 70 mph winds. They were designed for England conditions where they don't get thunderstorms like we do that bring wind and rain at the same time. The Hitts have already had some bows bend due to a downdraft thunderstorm. They learned that come the first of May, they vent the tunnels, using ladders to push the plastic as high as they can. They make sure to roll the plastic up inside so it doesn't wrinkle and collect water running off the top – this would soon weigh it down and cause bending. They are not designed to take a snow load. They must be uncovered in winter. They can be used with sidewalls and end walls but they still are not designed for overwintering crops. Like any greenhouse, there will be a lot of water running off so you need to prepare for this when preparing the site. Make sure that when the water pours down the sides, it can run off between the rows and not across them. Alex digs a little trench in the leg row and seeds it with a cover crop to direct water out into the field. Would they do it again, knowing what they know now? Alex said they would. For them, it was a specific answer to a specific problem, and tomatoes were 16% of their business so it was worth solving that problem. For people who want to grow lettuce all season, a Haygrove is not for them. They should just get a high tunnel. The Haygroves are pretty labor-intensive, and it takes a couple of years to learn how to best manage them. Shade Although season extension usually brings to mind an image of protecting plants from the cold, modifying temperatures in mid-summer can also be important. Shade over a bed can create a cool microclimate that will help prevent bolting and bitterness in heat-sensitive crops such as lettuce and spinach, make it possible to grow warm-weather crops in areas with very hot summers, and hasten germination of cool-weather fall crops. Some growers provide cooling shade by growing vines such as gourds on cattle panels or similar frames placed over the beds. Shade fabrics, available from greenhouse- and garden-supply companies, can be fastened over hoops in summer to lower soil temperatures and protect crops from wind damage, sunscald, and drying. Placing plants under 30 to 50% shade in midsummer can lower the leaf temperature by 10° F or more. Commercial shade fabrics are differentiated by how much sunlight they block. For vegetables like tomatoes and peppers, use 30% shade cloth in areas with very hot summers. For lettuce, spinach, and cole crops, use 47% in hot areas, 30% in northern or coastal climates. Use 63% for shade-loving plants. (The maximum shade density — 80% — is often used over patios and decks to cool people as well as plants). Shade houses can also provide frost protection for perennials and herbs during winter. Temperatures inside can be as much as 20° F higher than outdoors.

Fruit Trees & Vine

A fruit tree is a tree which bears fruit that is consumed or used by animals and humans — all trees that are flowering plants produce fruit, which are the ripened ovaries of flowers containing one or more seeds. In horticultural usage, the term "fruit tree" is limited to those that provide fruit for human food. Types of fruits are described and defined elsewhere (see Fruit), but would include "fruit" in a culinary sense, as well as some nut-bearing trees, such as walnuts.

The scientific study and the cultivation of fruits is called pomology, which divides fruits into groups based on plant morphology and anatomy. Some of those groups are pome fruits, which include apples and pears, and stone fruits, which include peaches/nectarines, almonds, apricots, plums and cherries.

The cultivation of fruit trees is one strategy that we feel can fight hunger on a long-term, sustainable basis. Though the importance of raising fruit trees has been underestimated by development communities, it ought to be a major element in any development scheme.

Fruits and nuts, when eaten in the right amounts and combinations, are capable of providing all the necessary nutrition that the body needs, including protein, carbohydrates, fats, vitamins, minerals, oils, and sugars. They also provide great enjoyment from the variety of tastes and sweetness that other crops don't provide. With proper selection of fruit tree species, you can have different kinds of fruit all year round. Once fruit trees are established, very little labor is required to maintain them and they continue to produce for many years. They will produce food even during difficult times when other garden produce may be hard to obtain. Fruit trees can also provide other benefits that include lumber, poles, medicine, income, shade, firewood, ornamental value, soil improvement, reforestation and protection of the environment.

Certain plants always grow as vines, while a few grow as vines only part of the time. For instance, poison ivy and bittersweet can grow as low shrubs when support is not available, but will become vines when support is available.

A vine displays a growth form based on very long stems. This has two purposes. A vine may use rock exposures, other plants, or other supports for growth rather than investing energy in a lot of supportive tissue, enabling the plant to reach sunlight with a minimum investment of energy. This has been a highly successful growth form for plants such as kudzu and Japanese honeysuckle, both of which are invasive exotics in parts of North America. There are some tropical vines that develop skototropism, and grow away from the light, a type of negative phototropism. Growth away from light allows the vine to reach a tree trunk, which it can then climb to brighter regions.

The vine growth form may also enable plants to colonize large areas quickly, even without climbing high. This is the case with periwinkle and ground ivy. It is also an adaptation to life in areas where small patches of fertile soil are adjacent to exposed areas with more sunlight but little or no soil. A vine can root in the soil but have most of its leaves in the brighter, exposed area, getting the best of both environments.

The evolution of a climbing habit has been implicated as a key innovation associated with the evolutionary success and diversification of a number of taxonomic groups of plants. It has evolved independently in several plant families, using many different climbing methods, such as:

- twining the stem around a support (e.g., morning glories, Ipomoea species)
- by way of adventitious, clinging roots (e.g., ivy, Hedera species)
- with twining petioles (e.g., Clematis species)
- using tendrils, which can be specialized shoots (Vitaceae), leaves (Bignoniaceae), or even inflorescences (Passiflora)
- using tendrils which also produce adhesive pads at the end that attach themselves quite strongly to the support (Parthenocissus)
- using thorns (e.g. climbing rose) or other hooked structures, such as hooked branches (e.g. Artabotrys hexapetalus)

The climbing fetterbush (Pieris phillyreifolia) is a woody shrub-vine which climbs without clinging roots, tendrils, or thorns. It directs its stem into a crevice in the bark of fibrous barked trees (such as bald cypress) where the stem adopts a flattened profile and grows up the tree underneath the host tree's outer bark. The fetterbush then sends out branches that emerge near the top of the tree.

Most vines are flowering plants. These may be divided into woody vines or lianas, such as wisteria, kiwifruit, and common ivy, and herbaceous (nonwoody) vines, such as morning glory.

One odd group of vining plants is the fern genus Lygodium, called climbing ferns.[10] The stem does not climb, but rather the fronds (leaves) do. The fronds unroll from the tip, and theoretically never stop growing; they can form thickets as they unroll over other plants, rockfaces, and fences.

Raising Chickens for Eggs

If you plan to start or have started raising chickens for egg production, you need to understand flock production capabilities. You need to know how to gauge the number of eggs your flock can produce and be aware of the variables that affect egg production. You should be able to identify which hens are laying and determine why your hens are not laying. By having a firm grasp of these factors, you will help ensure the success of your flock.

Production Expectations And Variables Affecting Production

A hen can lay only one egg in a day and will have some days when it does not lay an egg at all. The reasons for this laying schedule relate to the hen reproductive system. A hen's body begins forming an egg shortly after the previous egg is laid, and it takes 26 hours for an egg to form fully. So a hen will lay later and later each day. Because a hen's reproductive system is sensitive to light exposure, eventually the hen will lay too late in a day for its body to begin forming a new egg. The hen will then skip a day or more before laying again. See the related article discussing the reproductive tract of a chicken for more information on the specifics of egg production.

Also, hens in a flock do not all begin to lay on exactly the same day, nor do they continue laying for the same length of time.

The length of time that a flock will produce eggs varies as well. Many home flocks produce eggs on and off for three to four years. Each year, the level of egg production is lower than the previous year. Also, egg size increases and shell quality decreases each year.

Both the number of eggs you can get from a flock and the number of years a flock will produce eggs depend on several variables, including the following factors:
- Breed
- Management of pullets prior to lay
- Light management
- Nutrition
- Space allowances

Breed

Some commercial breeds of chickens have been developed specifically for egg production. The commercial White Leghorn is used in large egg production complexes, but these birds typically do not produce well in home flocks. They are simply too flighty. Moreover, they lay white-shelled eggs. People purchasing eggs from small flocks often prefer to buy brown-shelled eggs, even though no nutritional differences exist between brown-shelled eggs and white-shelled eggs.

Breeding companies also have developed commercial layers for brown-shelled egg production, with some bred specifically for pasture poultry production. In addition, many hatcheries sell what are called sex-link crosses. These specific crosses allow the hatchery to sex the chicks at hatch based on feather color. As a result, the number of sexing errors is reduced, so you are less likely to get an unwanted rooster.

Some people like having a flock composed of different breeds. Such a flock can produce eggs having a selection of shell colors. Many dual-purpose breeds, such as Plymouth Rocks and Rhode Island Reds, lay eggs with light brown shells. Maran hens lay eggs with dark, chocolate-colored shells, which have become popular lately. The Araucana is a South American breed that has feather tufts around the face and no tail and lays eggs having light blue shells. By crossing Araucanas with other breeds, breeders have produced "Easter Egger" hens that lay eggs with light blue, green, or pink shells. The chickens produced from these crosses have beards and muffs rather than the tufts seen on Araucanas, and they have tails. If bred to the purebred standards, such a cross will result in an Ameraucana, which lays eggs having blue-green shells.

Obviously, you can choose from several breeds. When making your decision about which breed or breeds to raise, keep in mind that commercial-type hens may give you a higher level of production initially, but other breeds tend to lay for more years. For additional assistance in deciding which breed to choose, see the related article on which chicken breed is best for a small or backyard flock.

Pullet management

It is important to manage pullets correctly, especially in the areas of nutrition and light management, because correct management will affect the level and quality of egg production once the birds start to lay. If the pullets come into production too early, they may have problems with prolapse, which can cause health problems across the flock. Also, the hens may lay smaller eggs throughout the production cycle.

When raising pullets from day-old chicks, brood the chicks as you would any other type of chick. See the related article on brooding poultry hatchlings for information about the basic care of chicks. For future laying flocks, keep in mind that light management is important from brooding through all laying periods.

If you purchase pullets ready-to-lay, you should ask how the pullets were raised with regard to nutrition and light management so that you can adjust your subsequent management of the flock accordingly. For example, you may have to delay light stimulation if the hens are too small.

Light management for year-round production

Chickens are called long-season breeders, meaning that they come into production as days become longer. That is, they start producing eggs when there are more hours of light per day. Typically, day-old chicks are kept on 23 to 24 hours of light per day for the first few days to make sure that they are able to find food and water, especially water. After that time period, you should reduce the number of hours of light per day. If you are raising the birds indoors, you can give them just 8 hours of light per day. If you are exposing them to outdoor conditions, you are limited by the number of hours of light per day in your area, of course. When the pullets are ready to start laying, slowly increase the light exposure until they are exposed to about 14 hours of light per day. This exposure should stimulate the flock to come into lay. To keep the flock in lay year-round, you will need to maintain a schedule of at least 14 hours of light per day. You can increase the amount of light slowly to 16 hours per day late in the egg production cycle to help keep the flock in production. For most flock owners, this strategy involves providing supplemental lighting. Using a light with a stop/start timer, you can cause the light to come on early in the morning before sunrise and in the evening before sunset to ensure that the length of light exposure for the flock totals 14 to 16 hours. Also, you can get a light sensor so that the light bulb does not come on when natural daylight is available. By using such a device, you minimize your electricity use. The supplemental light you provide does not have to be overly bright. A typical 60-watt incandescent light bulb works fine for a small laying flock. For a discussion of other light choices, watch the recording of the webinar Lighting for Small and Backyard Flocks by Dr. Michael Darre from the University of Connecticut.

Nutrition

Chickens of any type and age require a complete, balanced diet. Feed mills assemble the available ingredients in combinations that provide all the nutrients needed by a flock in one package. Some producers mix complete feeds with cheaper scratch grains, but doing so dilutes the levels of nutrients the chickens are receiving, and nutrient deficiencies can occur. Nutrient deficiencies can adversely affect the growth of pullets and the level of production of hens.

It is also important to feed the specific feed tailored for the type and age of the chickens you have. For example, do not feed a "meat-maker" type diet to growing pullets or laying hens as it will not meet their nutritional needs. Likewise, do not feed a layer diet to growing chickens. The diet of a laying hen is high in calcium, which is needed for the production of eggshells. This level of calcium, however, is harmful to non-laying chickens.

Some hens have a higher need for calcium than others. It is always good to have an additional source of calcium available. Oystershell, usually available in feed stores, is an excellent calcium supplement for a laying flock.

Space allowances

To produce effectively, laying hens must have adequate space. The amount of floor space required by a flock depends on the size of the chickens (which is related to the breed of chicken chosen) and the type of housing used. A minimum of 1.5 square feet per hen is recommended, with 2 square feet per hen being the most commonly used space allowance. Larger allowances are required for some of the larger breeds.

To make use of the entire housing facility, you can incorporate perches. The hens will sleep on the perches at night, keeping them off the floor. The use of perches also helps concentrate much of the manure in a single location for easier cleaning of the poultry house. Moreover, chickens have a desire to perch, so providing for this behavior contributes to animal welfare. For more information, read the related article on perches. If you provide outdoor space for your chickens, the amount of outdoor space needed depends on the quality of the space. If your goal is to maintain a pasture, you will require more area than you would need if simply providing outdoor access for a small backyard flock. An allowance of 2 square feet per hen typically is recommended for simple outdoor access. If you do provide your flock with outdoor access, be aware of predator possibilities from both the ground and the air, and provide the hens with the protection they require.

Identification Of Laying Hens

To determine which of your hens are laying, it is important to know more about the type of hens you have. For many breeds, hens that are laying eggs have large, bright red combs and wattles. For other breeds, the combs and wattles are normal color during the laying period but fade after the laying period. For hens with yellow pigment in the skin, such as Rhode Island Reds and Plymouth Rocks, the level of pigmentation is a good indication of where the hens are in the production cycle. Hens lose the yellow pigment in a specific order. The color fades first from the vent; then the face (beak, eye ring, and earlobe); and then the feet (shanks, toes, and hock). An additional method for identifying laying hens involves evaluating the level of fat in the abdomen and the abdominal capacity as measured by the distances between the pubic bones (abdominal width) and between the pubic bones and the tip of the keel, or breastbone (abdominal depth). The lower the level of fat and the larger the abdominal capacity, the more likely the hen is to be laying.

Reasons Hens Stop Laying

Many factors can affect egg production, with health (before and after lay) being one of the most significant. If your hens stop laying, you may be able to identify the source of the problem by asking the following questions:
• Have the hens been laying for 10 months or more? Your hens may just be at the end of their laying cycle. If so, they will stop production, go through a molt (loss of feathers), take a break, and start laying again. If your hens have been laying for less than 10 months, something else may be causing their lack of production.
• Are the hens receiving enough fresh, clean water? The hens will not eat if they cannot drink, so make sure that your watering system is functioning correctly. Keeping a watering system operational can be a challenge in the winter when the water may freeze. You can purchase waterers that have heaters attached to keep the water from freezing. Otherwise, you will have to break up any frozen water on a regular basis. Problems can occur in summer as well. Summertime high temperatures can make the water so warm that the chickens will not drink enough to meet their increased needs. For more information, refer to the related article on the water requirements of poultry.
• Are the hens eating enough of the right feed? Feeding the wrong feed, diluting feed with scratch grains, or limiting the amount of feed available can result in your hens having a nutritional deficiency, causing them to molt and go out of production. When hens have a nutritional deficiency, it is common to see feather pecking as well as a loss of egg production.

• Are the hens getting enough hours of light per day? Decreases in the number of hours of light per day typically will put a flock out of production. For this reason, many flocks that are not provided with supplemental light go out of production during the fall and winter months.
• Do the hens have parasites? Various internal parasites and external parasites can infest poultry flocks and stress the hens. Heavy infestations of internal parasites can result in serious damage to the digestive tract and reduce hen performance. Heavy infestations of mites can cause anemia in the hens, also adversely affecting their performance.
• Did any issues with eggshell quality precede the stop in egg production? Several diseases can result in abnormal eggshells.
• Have there been any health issues within the flock? A flock that has been sick will not perform as well as a flock that has not gone through a disease challenge.

Handling Health Issues

Because chickens have fast metabolisms and high body temperatures, any sickness they contract will progress quickly. If you notice listlessness, coughing, sneezing, standoffishness, or other unusual behavior, isolate the hen in question and stir a powdered vitamin-electrolyte product into her water, refreshing it twice a day. You can seek advice from your county agricultural agent, local extension office, or, if the chicken's health seems to decline rapidly, a qualified farm vet. Also, keep an eye out for similar signs among the rest of the brood.

Don't be alarmed, however, if your chickens start to look bedraggled near the end of summer. The shortening days signal that it's time to molt, and the birds begin shedding old feathers and growing new ones for insulation. Laying will wane and may stop altogether, as the birds direct their energy and nutrients toward the new growth. Lastly, a note about broody hens: Occasionally, one will want to sit on a clutch of eggs until they hatch. Removing her from the eggs and moving her nesting box can help snap her out of the broodiness, though it may take weeks.

Reaping the Benefits

The hard work of chicken-keeping pays off in the form of freshly laid eggs. Pasture-raised birds can lay eggs with a third less cholesterol, a quarter less saturated fat, and two-thirds more vitamin A than typical commercially produced eggs. Plus, homegrown ones can be stored, unrefrigerated, for several weeks, as long as they're unwashed and unfertilized. (If there's a rooster about, refrigerate the eggs within 36 hours.) Before consuming an egg, wash it under hot water to remove the protective cuticle; the water must be significantly warmer than the egg because of the porous nature of shells. In the six years since my husband returned from the feed store with those five fateful chicks, our operation has grown to include partner farms and some 20,000 birds. While my role these days is mostly limited to washing eggs, I know the answer to that age-old question: The chicken always comes first.

Raising Chickens for Meats

If you're interested in raising chickens for meat, not eggs, you'll need to learn a few things and prepare your chicken raising a little bit differently. There are some additional steps to consider, including the slaughtering, processing, or butchering the birds when they are fully grown to market size. Chickens raised for meat are commonly called "meat birds" and are usually a different breed from laying hens.

Should You Raise Meat Birds?
Before you get the chicks, consider whether you want to raise meat birds. They're very different from laying hens. You'll have a lot (usually 50 or more, although you could just raise a few) of fast-growing birds. This also means you'll have a lot of poop. It's important to ask yourself (and your family) if you can handle saying goodbye in six to eight short weeks. Whether you slaughter them on-farm or take them to be processed, you will need to face this reality. Also note, it's cruel to meat birds to let them live longer than a few months as they are heavy-breasted and can die of heart failure if they grow too big.

How to Choose a Meat Bird Breed
Meat birds are a separate breed from laying hens. Although a hundred years ago laying hens were truly dual-purpose, meaning most people kept a flock of hens and roosters and killed older birds as needed for meat, older chickens tend to be tough and stringy, better for stew or soup than a roast chicken like you eat today.

Cornish Rocks, which are a cross between a Cornish and a White Rock, are the typical meat bird breed and used in factory farms all over the United States and on many small family farm operations as well (both pastured and conventional). They are extremely efficient converters of feed to muscle. Other breeds more suited to pasture are also becoming available.

How to Choose a Coop for Meat Birds

You will need a coop for your chickens, just like for your laying hens. Coops for meat birds are often larger so that you can raise 50, 100, or more birds at a time. Many people raise meat birds just during the summer season. This way they can often be in more temporary shelters such as hoop houses or tarps. You will need to make sure your birds have protection from the rain and the wind. They don't need roosts because meat birds don't like to roost. If you're pasturing your chickens, you will want to have a movable coop or use a day ranging method.

How to Start From Day-Old Chicks

Most likely, you will buy your chickens as day-old chicks from a hatchery or feed store. Baby chicks require a bit of specialized care. They need a brooder area and a heat lamp to keep them warm. They need their brooder temperature monitored closely and they need to be prevented from developing issues like pasting up.

Raising Meat Birds on Pasture

You can keep your chickens in a coop with just a small run attached, but meat birds raised on pasture tend to produce meat that is higher in omega-3s and the birds are just happier.

Processing Chickens on the Farm

When your birds have grown to full size, typically 5–7 pounds depending on whether you're raising broilers or roasters, it's time to process them into chickens for the freezer. You can do this on-farm or you can find a poultry processor and transport the birds to the site to be slaughtered and processed. If you plan to sell your birds at a store or farmers market, you will need to have them slaughtered at a USDA-approved facility. Some states have mobile facilities, making the location more flexible.

Meat chicken breeds

Cornish, Plymouth Rock and New Hampshire breeds are the most economical meat strains. These crosses feather rapidly and mature early and have the most economical conversion of feed to poultry meat.

Some flock owners use White or Barred Plymouth Rocks, Rhode Island Reds and New Hampshires for meat. These breeds generally don't grow as rapidly as the crosses and take more feed per pound of weight gained. Leghorn males don't make good meat birds and are unprofitable even if you receive day-old chicks.

The various classes of chicken meat birds are raised from the same commercial strains.

• Broilers or fryers: birds slaughtered at 7 to 9 weeks of age when they weigh 3 to 5 pounds and dress a 2 ½ to 4-pound carcass
• Cornish game hen: birds slaughtered at 5 weeks of age
• Roasters: birds grown out to 12 weeks or longer
• Capons: male birds neutered at 3 to 4 weeks and marketed after 18 weeks

Meat-type chicks are usually purchased on a straight-run basis (males and females mixed).

When ordering chicks

• Hatcheries in Minnesota and around the country can be found online.
• Plan their arrival around their departure. Cornish cross broilers (most commonly raised) need only six to eight weeks to reach a market carcass weight of four to six pounds. Other breeds that grow slower may take 10 to 12 weeks.
• You can order cockerels (males), pullets (females) or a straight run (mixed batch). Cockerels are a little more expensive but grow faster. They may weigh one pound more than pullets at processing, at the same age.
• You will need to arrange processing well in advance.
• If you grow birds for your own consumption within the limits of a town or city, check the local government ordinances prior to processing the birds in your backyard, as this is not typically allowed.
• Consider having the birds vaccinated at the hatchery against coccidiosis. It is cheap to do so, and then you will very likely have healthy birds throughout their short growing period. This vaccine will help give the birds protection against a very common and costly poultry disease. Doing so can then give you the option of using non-medicated feed throughout the production period.

Housing

When raising chickens for meat, they will need about ½ sq. foot of space per bird. When you are starting out. This isn't a lot of space – but they are going to grow rapidly.
Until they are 2 weeks of age, then you will need to provide 1 sq. foot of space per bird until processing time. If you can provide them with 2-3 sq. feet of space per bird, they will have much more room to move around, and the floor will stay cleaner and drier.
You have a few options in terms of how you set up your chicken housing. Since meat birds grow out so quickly, you have more housing options than you would if you were raising a flock of laying hens to be kept throughout the entire season.

A popular option for raising meat birds is in chicken tractors. Chicken tractors are portable chicken housing that can be moved around on pasture in your back lawn. You move the tractor every day, or every few days, depending on your stocking density, and the chickens have access to fresh grass each and every day.

A major benefit of raising your chickens in a chicken tractor is that they are always on fresh ground so there is no need to control for parasites or other illnesses. Your feed bills will be dramatically lower, since the birds can forage, and the upfront costs to build a chicken tractor tend to be a lot less than those involved in building or buying a fixed-location coop.

Basically, chicken tractors just need to offer shelter from the wind and rain. You can use tarps, even, to suit this purpose, as long as your birds are protected against predators. However, if you want to build a permanent structure, a coop is a great option, too. Just make sure you give your birds some access to outsides pace.

As meat chickens, they won't need a ton of room to roam, but remember that giving your chickens access to green grass and sunshine will reduce your likelihood of disease – plus, it will reduce the frequency with which you need to clean the coop!

If you use litter for the flooring, it should be maintained to a depth that is dry and fluffy, somewhere around 6 inches deep. If you have them on fresh grass, moving their pen daily will keep them healthy.

When they are on excessively wet, caked, or dirty litter or ground, they can develop breast blisters. Breast blisters can easily lead to a loss of that meat.

You also need to make sure your meat chickens have a good roost bar system. They will need a place to sleep up and off the ground (disregard this, of course, if your chickens are in a chicken tractor) and these roost bars shouldn't be quite as high as they might be if you were raising laying chickens.

The reason for this is that as they grow larger, your chickens' legs won't be quite as strong as they used to be – you don't want them to break their legs hopping off the roost. You won't need nest boxes in your chicken coop, as you would with a flock of laying hens, but you will need to make sure your coop is properly ventilated.

This can prevent things like chicken lice and chicken mites and it will also keep your chickens cooler in the hot days of summer. Make sure your coop offers protection from predators, especially when your chickens are young.

Even chickens for meat like dust baths to stay cool, but when they get larger, they are not always able to kick enough dust on themselves to cool off. Providing a sprinkler, plenty of water and shade will keep them cool. In addition to a dust bath, you should make sure that your chickens are given ample shade.

This is one of the major benefits of a chicken tractor – on a particularly hot day, you can move them to a shady location under a tree, but if It's cold, you can put them out in the sunlight.

Food and Water

Having constant access to clean water is a key to bird health. You will want to refresh their water on a regular basis. This means at least once a day, twice when it's hotter outside.

We have added pennies to the watering bases to encourage them to peck at the shiny objects and drink more. Placing a light yard sprinkler by the waterers will give them a "shower" as they get a drink. This will help cool them off.

As for what to feed meat chickens, a higher protein food is needed. The feed that layer chickens will get is lower in protein, usually 16-18%. Roaster chickens for meat, or broilers need a feed that is 21% protein to gain weight properly.

While you can easily make your own DIY chicken feeds to save money, it's a good idea to use soybeans in your meat bird feed, as these tend to fatten up your birds the quickest. Feeders must also be large enough to supply the flock's needs for at least a full day. They will need to consume feed in order to gain weight. You will want to plan for about 10 to 15 lbs of feed per bird from chick to market weight.

You can get them to consume more feed just by going out there 3-4 times a day, and adding as little as ½ cup of feed to the feeder, and they will think they need to eat again since you are adding more food.

This is one of the tricks in how to make a broiler chicken grow faster. You need to provide plenty of feed at all times. Placing a light over the food at night will also keep them eating more. If the food is there, and they can see it, they will eat.

A good option, if you are pressed for time, is just to provide free access to feed. Many farmers do this to encourage their birds to eat.

Remember, if your chickens don't have water, they will be less likely to eat – you need to make sure they have plenty of fresh, clean water to help them prevent dehydration and to encourage them to eat on the hottest days of summer.

I recommend putting the water on the other side of the chicken pen. Some broiler breeds of chicken have a tendency to be extremely lazy – if you have food and water on two opposite ends of the pen, this will force them to get up every now and then. This way, they will form lean muscle instead of tons of fat -which is what you want in a meat chicken.

The Markham farm chicken plucker

Public Domain Thresher Designs

Threshing Machine

A threshing machine or a thresher is a piece of farm equipment that threshes grain, that is, it removes the seeds from the stalks and husks. It does so by beating the plant to make the seeds fall out.

Before such machines were developed, threshing was done by hand with flails: such hand threshing was very laborious and time-consuming, taking about one-quarter of agricultural labour by the 18th century. Mechanization of this process removed a substantial amount of drudgery from farm labour. The first threshing machine was invented circa 1786 by the Scottish engineer Andrew Meikle, and the subsequent adoption of such machines was one of the earlier examples of the mechanization of agriculture. During the 19th century, threshers and mechanical reapers and reaper-binders gradually became widespread and made grain production much less laborious. Michael Stirling is said to have invented a rotary threshing machine in 1758 which for forty years was used to process all the corn on his farm at Gateside. No published works have yet been found, but his son William made a sworn statement to his minister to this fact. He also gave him the details of his father's death in 1796.

Separate reaper-binders and threshers have largely been replaced by machines that combine all of their functions, that is combine harvesters or combines. However, the simpler machines remain important as appropriate technology in low-capital farming contexts, both in developing countries and in developed countries on small farms that strive for especially high levels of self-sufficiency. For example, pedal-powered threshers are a low-cost option, and some Amish sects use horse-drawn binders and old-style threshers.

As the verb thresh is cognate with the verb thrash (and synonymous in the grain-beating sense), the names thrashing machine and thrasher are (less common) alternate forms. Early threshing machines were hand-fed and horse-powered. Some were housed in a specially constructed building, a gin gang, which would be attached to a threshing barn. They were small by today's standards and were about the size of an upright piano. Later machines were steam-powered, driven by a portable engine or traction engine. Isaiah Jennings, a skilled inventor, created a small thresher that does not harm the straw in the process. In 1834, John Avery and Hiram Abial Pitts devised significant improvements to a machine that automatically threshes and separates grain from the chaff, freeing farmers from a slow and laborious process. Avery and Pitts were granted United States patent #542 on December 29, 1837.

John Ridley, an Australian inventor, also developed a threshing machine in South Australia in 1843.

Preserving your Harvest

Storing your harvest is a great way to deal with gluts (a surplus of one vegetable) and months when little is growing. There are many ways to store your vegetables; these include drying, freezing and preserving.

1. Fresh Storage

Given the right conditions, a number of garden-fresh crops last a surprisingly long time in storage — no processing necessary. Carrots, cabbage, kohlrabi, onions, potatoes, winter squash and apples fall into this long-lived category, lasting anywhere from one to six months, depending on the crop.

Proper storage, however, is essential: Some produce does best kept under cold, humid conditions — think root cellar (an example of which is pictured above) — while others last longer in cool, dry environments, such as a garage. All need protection from nibbling rodents. Certain varieties keep better than others, and when evaluating candidates for storage, pick good fruits and vegetables free of blemishes, nicks and bruises. No bad apples allowed!

2. Freezing

The freezer is a marvelous modern invention, capable of putting foods in the no-rot zone for months. Proper freezing of fruits and vegetables tends to maintain their color, flavor and nutrients better than other preservation methods. Don't just toss those fresh-picked beans in the freezer, though: To kill bacteria and inhibit the action of flavor-altering enzymes, vegetables must undergo a quick boiling water bath treatment called blanching before they're frozen.

Fruit prep for freezing ranges from a simple wash/drain/pack to blanching and packing in sugar syrup. Caution: Avoid freezing raw produce with a high water content, such as salad greens, cucumbers, cabbage and watermelons. Upon thawing, you're likely to end up with a soggy mess on your hands.

3. Freezer Jam

If you're just getting into preservation, making freezer jam with berries is a fun and easy way to start. Quicker than preparing cooked jams, this method tends to yield fresher-tasting and more vividly hued jams with a softer set. All you need to begin: sugar, pectin, ripe berries, freezer containers and some basic kitchen equipment. You can find instructions for making various freezer jams and jellies included with boxes of pectin. For those trying to cut back on sugar, there is pectin for making low or no-sugar jams, too. These jams will last up to a year in your freezer.

4. Pesto

Say "pesto," and most people think of that saucy Italian mingling of basil, pine nuts, garlic, Parmesan cheese and olive oil. But when your garden has become a lush kale jungle or you're faced with a whiskey barrel crammed with use-it-or-lose-it cilantro, it's time to step outside the classic pesto box. Other leafy greens and herbs can substitute for basil in pesto recipes with tasty results, including kale, arugula, spinach, cilantro and parsley. With a good blender, food processor, or even a mortar and pestle, making several fresh pesto batches — basil or otherwise — takes no time at all. Pesto freezes well for up to a year; try freezing it in ice cube trays to add to soups or vegetables, or in larger containers for speedy, fragrant pasta dishes. Hint: Save money by switching out pine nuts for walnuts.

5. Water Bath Canning

A cupboard gleaming with harvest-packed glass canning jars warms a farmer's heart. Whether filled with vibrant cooked jams, tart applesauce or sugar-sweetened pears, they seem to epitomize self-reliance — and unlike freezer-preserved edibles, they'll weather an extended power outage just fine. Proper preparation and processing of berries and other high-acid fruits, such as apples and pears, in a boiling water bath canner will make them safe to store on the shelf for up to a year.

To get started, you'll need at minimum a large canning kettle with rack and lid, jar-lifter, wide-mouth funnel, jars, and self-sealing lids/rings. Tip: Try asking an experienced canner friend or relative to give you a lesson, or see if your local community college offers classes. You'll stress less and have more fun.

6. Salsa

If your garden glows with an abundance of tomatoes, tomatillos and peppers, then salsa-making should step to the head of your harvest-time task list — no dance lessons required. As a bonus, many salsa recipes incorporate surplus cilantro, onions and garlic, too. First, I suggest putting on some peppy music: Salsa preparation requires plenty of chopping (or food processing) of ingredients — tomatoes for red salsa, tomatillos for green, and pepper varieties based on the recipe and how hot you like your salsa.

Next, add any recommended seasonings plus an acidifying agent, such as vinegar or lemon juice, because tomatoes vary in their spoiler-fighting acid content. After some simmering, you'll have a vibrant, spicy concoction to ladle into jars and process in your water bath canner. Salsa can also be frozen in containers for several months.

7. Pickling

What would a burger be without pickles or a hotdog without relish? Once you've embraced the water bath canner, it's not difficult to harness vinegar's preservative powers to transform surplus veggies and fruits into shelf-safe pickles, relishes and spicy chutneys (or pickled shallots, pictured above). Unlike fermented vegetables — No. 13 on our list — fresh-pack pickled foods don't require a long fermentation period before canning, though their taste does benefit from curing for about four to six weeks. Pickling instructions vary with the food being pickled, but you'll find plenty of recipes to play with, from plain old dill pickles to green tomato relish. Suggestion: Still got monstrous zucchini lurking in the garden? Make a dent in their population with zucchini pickles or sweet relish.

8. Pressure Canning

Unlike acidic fruits or vinegar-pickled veggies, low-acid foods, such as corn, beans and carrots, require careful processing in a pressure canner to make them safe for long-term shelf storage. This is because the bacteria that cause botulism live in the soil and on our produce, and at room temperature, they can generate toxins in anaerobic, low-acid environments. Boosting the temperature of foods to 240 degrees F and higher — and keeping it there for a specified amount of time — destroys these pathogens.

While today's weighted-gauge and dial-gauge pressure canners are safer to operate than older models, they can still be pretty complicated and intimidating. Your best bet for a safe and successful pressure canning experience is to find a class or an experienced canner to show you the ropes.

9. Dried Herbs

To treat yourself to herbs' aromas and tastes all year, tie clean, unsprayed herb stems into bundles with floss or string. Then, hang them upside down in a dark, dry, ventilated area, such as a pantry or closet with the door ajar. If you're worried about dust or discarded leaves, enclose them in perforated paper bags. Wait a few weeks or so — drying time will depend on the herb and your home's humidity — and voilà, you'll have your own dried seasonings or the makings of a soothing cup of tea. Tightly bottled and stored in a dark place, they keep for up to a year. Herbs that retain their flavor well when dried include oregano, rosemary, tarragon, thyme and mint. Dry hot peppers and garlic this way, too.

10. Vine-Dried Beans

High in fiber and protein, beans have a well-deserved reputation as a super-healthy food. They also happen to be one of the easiest crops to grow — not to mention one of the simplest to preserve by drying. Given a stretch of sunny weather, you can just let the mature pods wither on the vine until they're brown and the seeds rattle inside, and then get busy shelling. If rain threatens before this stage, pick them anyway. You can finish drying them inside, either on trays or in your oven/dehydrator; just be sure to shell them first.

To kill pests, shelled beans should be placed in bags in the freezer for two days before storage. If you grow heirloom varieties, don't forget to save some for planting next year, too.

11. Dehydrating

The oldest preservation method around, dehydration prevents and delays spoilage by removing most of a food's water content. The result? Incredible shrinking grapes, tomatoes, berries and more, compact and light enough to pack on a hike or a trip. With proper pretreatment, such as blanching veggies or immersing apples in a solution to prevent browning, you can safely dehydrate most fruits and vegetables. While it's possible to dry them in an oven or even in the sun, you'll get more dependable results using an electric dehydrator. These nifty appliances with tiered drying trays provide a controlled temperature—hot enough to dry the food without cooking it—and contain a fan to circulate air and remove moisture. Keep in mind that drying times can vary according to the food, how it's prepared and other factors. Consult the manual that comes with your dehydrator.

12. Flavored Vinegar

Usually made from water and a mix of fermented corn, apples or wine, vinegar's spoiler-busting acetic acid content make it a crucial ingredient in canned salsa, tomato and pickled vegetable recipes. This tart preservative can also be infused with cleaned herbs, spices and/or fruits to make flavored vinegars (such as pesto vinegar, shown above). The process usually involves letting a combination of the vinegar and flavoring ingredients steep in canning jars or a covered glass bowl in a cool, dark place for three to four weeks. After straining, the vinegar can be bottled and kept in the fridge or processed in a water-bath canner before storing. Use flavored vinegars to liven up salads or steamed veggies.

13. Fermenting

If you're scratching your head over oodles of cucumbers or cabbage, you may want to try your hand at fermentation, another ancient means of preservation. Submerging cucumbers along with spices in a salt water solution, then storing them in a cool place for a few weeks, leads to the formation of lactic acid, a preservative. The same thing happens when you layer shredded cabbage with salt, pack it into a pickling container, then cover it with brine and let it ferment for about three to six weeks. At the end of this period, you'll have tangy pickles and sauerkraut to jar up and process in a water-bath canner. Bonus: Traditionally fermented foods are rich in good-for-your-gut probiotics.

14. Alcohol Infusions

Preservation doesn't get much easier than this: Pack a sterilized canning jar with washed cherries, plums or other fruit, and pour in brandy, vodka or another high-proof alcohol to cover. Add some sugar to sweeten things up, screw on the lid and shake. That's it. Alcohol is an excellent preservative, but to be on the safe side, store your "drunken" fruit in the fridge as you wait several weeks for the flavors to meld. Then, try savoring them on desserts or even neat—in moderation, of course.

15. Craft Wine & Cider

I'll be up-front here: Making good wine and cider is an art form, one that requires quality ingredients, the right equipment and proper know-how. It also takes patience because the yeast-driven fermentation process does its sugar-to-alcohol magic over several months. But a well-crafted batch of sweet blackberry wine or effervescent apple cider is so worth the wait and a wonderful way to enjoy the surplus fruit of your labors. In fact, it's possible to make wine out of a surprising variety of produce, including apricots, carrots and rhubarb.

Wine-making equipment basics include primary and secondary fermenters, an air lock, food-grade tubing, and glass bottles and corks, plus bottle corker. For cidering, you'll also require a cider press — or access to one. You can find beginner's kits, essential ingredients, instruction books and invaluable advice at wine-making supply shops. Cheers!

Selling Your Produce

You have three main avenues to explore: dealing directly with restaurants, selling through farmers markets or creating your own market.
1. Create Your Own Market.
2. Work With Local Restaurants.
3. Sell at Farmers Markets.
4. Reach Out to Small Stores.
5. Certifications and Pricing.
6. Consumers: Support Local Growers.

Wholesale distributors play a key role in the food system by purchasing, aggregating, and transporting large volumes of food and distributing it to their customers. Distributors typically sell to one or more of the following groups:
• Independent and chain restaurants
• Institutions such as schools, universities, and hospitals
• Retailers and grocery stores
• Corporate cafeterias
• Catering companies

By nature, distributors are the intermediate step in the supply chain between farmers and consumers, making a distributor a door through which farmers and producers can reach many customers and market opportunities. As a farmer or producer, selling product to a distributor offers a variety of benefits:
• You can sell larger volumes of product more efficiently than selling directly to consumers.
• You can reduce the time and resources spent on marketing directly to consumers to focus more on production.
• You can access different markets to increase sales volume and brand presence.

Although the benefits may be great, working with a distributor comes with challenges and requirements that are very different from those of direct marketing. On-farm operations, product packaging, marketing, pricing, food safety certifications, and logistics are all components that need to be carefully considered to create a successful relationship with a distributor.

In this publication, we provide guidance to farmers and producers who are considering selling to distributors. This information will be useful to owners of small and medium-sized farms who have not previously sold their products to a distributor.

Is Distribution Right for You and Your Business?

As a farmer trying to grow your business, you would like to pursue working with a distributor. Consider the following questions as they relate to your operation to make sure that this is the best market outlet to pursue.

• What is your goal for working with a food distributor? How does this fit into your current and future business plan?
• What market are you targeting by working through a distributor? Do you have any preexisting relationships with these types of customers?
• What products do you want to sell through a distributor? When are these products available throughout the year? Will you be able to supply the distributor with consistent product throughout your growing season?
• What is your current production capacity? Will you be able to expand production if demand increases?
• What is your current labor capacity for product, postharvest operations, sales, transportation, and other aspects of production?
• What is your current sales volume per week? How will your other customers be affected if you work with a distributor? If you plan to sell to more than one distributor, will you have enough product?
• What is the target sales volume you hope to achieve by selling through a distributor? Is this enough to cover the additional input costs?
• How will you get your product to the distributor?

After examining your own goals and understanding what a distributor would expect from you as a vendor, you may have decided that working with a distributor is the right decision for you and your business. In this section, we describe how to connect with the buyers and decision-makers at the distributor.

When should you approach a distributor?

Typically, winter (December to February) is the best time to approach a distributor as they are planning with growers and customers for the upcoming season. They may be able to incorporate your products into their plan for the upcoming year, especially if you are prepared with your Annual Produce Availability Calendar (see Table 2). During the other months of the year, more crops are in season, so distributors are busy. You can still approach a distributor in the other months, but they may not have the time to work with you.

Who in the company should you talk to?

Ask to speak with the produce buyer or purchasing manager.

How can you make this connection?
• Visit the distributor's warehouse in person.
• Email or call the produce buyer or purchasing manager.
• Attend grower-buyer networking events. Ask your local Cooperative Extension agent if these events are already being held, or could be held, in your area.

How should you prepare for your meeting with the buyers?
• Bring your marketing materials and communication tools, samples or photos of your packed and unpacked product, and photos of your operation.
• Have several business or customer references for the distributor to contact, in case the distributor asks for references. Evidence that you are already selling to a retailer or restaurant gives the buyer confidence in your operation.

Putting It All Together

Agroecology — the science that underlies sustainable farming — integrates the conservation of biodiversity with the production of food. It promotes diversity which in turn sustains a farm's soil fertility, productivity and crop protection. Innovative approaches that make agriculture both more sustainable and more productive are flourishing around the world. While trade-offs between agricultural productivity and biodiversity seem stark, exciting opportunities for synergy arise when you adopt one or more of the following strategies:

- Modify your soil, water and vegetative resource management by limiting external inputs and emphasizing organic matter accumulation, nutrient recycling, conservation and diversity.

- Replace agrichemical applications with more resource-efficient methods of managing nutrients and pest populations.

- Mimic natural ecosystems by adopting cover crops, polycultures and agroforestry in diversified designs that include useful trees, shrubs and perennial grasses.

- Conserve such reserves of biodiversity as vegetationally rich hedgerows, forest patches and fallow fields.

- Develop habitat networks that connect farms with surrounding eco-systems, such as corridors that allow natural enemies and other beneficial biota to circulate into fields

Different farming systems and agricultural settings call for different combinations of those key strategies. In intensive, larger-scale cropping systems, eliminating pesticides and providing habitat diversity around field borders and in corridors are likely to contribute most substantially to biodiversity. On smaller-scale farms, organic management — with crop rotations and diversified polyculture designs — may be more appropriate and effective. Generalizing is impossible: Every farm has its own particular features, and its own particular promise.

Designing a Habitat Management Strategy
The most successful examples of ecologically based pest management sys- tems are those that have been derived and fine-tuned by farmers to fit their particular circumstances. To design an effective plan for successful habitat management, first gather as much information as you can. Make a list of the most economically damaging pests on your farm. For each pest, try to
find out:

- What are its food and habitat requirements?

- What factors influence its abundance?

- When does it enter the field and from where?.

Enhancing Biota and Improving Soil Health
Managing soil for improved health demands a long-term commitment to us- ing combinations of soil-enhancing practices. The strategies listed below can aid you in inhibiting pests, stimulating natural enemies and — by alleviating plant stress — fortifying crops' abilities to resist or compete with pests.

- Add plentiful amounts of organic materials from cover crops and other crop residues as well as from off-field sources like animal manures and composts. Because different organic materials have different effects on a soil's biological, physical and chemical properties, be sure to use a variety of sources. For example, well-decomposed compost may suppress crop diseases, but it does not enhance soil aggregation in the short run.
Dairy cow manure, on the other hand, rapidly stimulates soil aggregation.
- Keep soils covered with living vegetation and/or crop residue. Residue protects soils from moisture and temperature extremes. For example, residue allows earthworms to adjust gradually to decreasing tempera- tures, reducing their mortality. By enhancing rainfall infiltration, resi- due also provides more water for crops.

- Reduce tillage intensity. Excessive tillage destroys the food sources and micro-niches on which beneficial soil organisms depend. When you reduce your tillage and leave more residues on the soil surface, you create a more stable environment, slow the turnover of nutrients and encourage more diverse communities of decomposers.

Strategies for Enhancing Plant Diversity

As described, increasing above-ground biodiversity will enhance the natural defenses of your farming system. Use as many of these tools as possible to design a diverse landscape:

- Diversify enterprises by including more species of crops and livestock.

- Use legume-based crop rotations and mixed pastures.

- Intercrop or strip-crop annual crops where feasible.

- Mix varieties of the same crop.

- Use varieties that carry many genes — rather than just one or two — for tolerating a particular insect or disease.

- Emphasize open-pollinated crops over hybrids for their adaptability to local environments and greater genetic diversity.

- Grow cover crops in orchards, vineyards and crop fields.

- Leave strips of wild vegetation at field edges.

- Provide corridors for wildlife and beneficial insects.

- Practice agroforestry, combining trees or shrubs with crops or livestock to improve habitat continuity for natural enemies.

- Plant microclimate-modifying trees and native plants as windbreaks or hedgerows.
- Provide a source of water for birds and insects.

- Leave areas of the farm untouched as habitat for plant and animal diversity.

As you work toward improved soil health and pest management, don't concentrate on any one strategy to the exclusion of others. Instead, combine as many strategies as make sense on your farm. Nationwide, producers are finding that the triple strategies of good crop rotations, reduced tillage and routine use of cover crops impart many benefits. Adding other strategies
— such as animal manures and composts, improved nutrient management and compaction-minimizing techniques — provides even more

Rolling out your Strategy

Once you have a thorough knowledge of the characteristics and needs of key pests and natural enemies, you're ready to begin designing a habitat-management strategy specifically for your farm.

- Choose plants that offer multiple benefits — for example, ones that improve soil fertility, weed suppression and pest regulation — and that don't disrupt desirable farming practices.

- Avoid potential conflicts. In California, planting blackberries around vineyards boosts populations of grape leafhopper parasites but can also exacerbate populations of the blue-green sharpshooter that spreads the vinekilling Pierce's disease.

- In locating your selected plants and diversification designs over space and time, use the scale — field- or landscape-level — that is most consistent with your intended results.

- And, finally, keep it simple. Your plan should be easy and inexpensive to implement and maintain, and you should be able to modify it as your needs change or your results warrant.

In this book, we have presented ideas and principles for designing and implementing healthy, pest-resilient farming systems. We have explained why reincorporating complexity and diversity is the first step toward sus- tainable pest management. Finally, we have described the pillars of agro- ecosystem health:

- Fostering crop habitats that support beneficial fauna

- Developing soils rich in organic matter and microbial activity

Throughout, we have emphasized the advantages of polycultures over monocultures and, particularly, of reduced- or no-till perennial systems over intensive annual cropping schemes.